Dedication

This book is dedicated in memory of my father, Kenneth Owen Arnold, who always encouraged me to follow my dreams. When other adults discouraged me from entering the engineering field, he told me, "If you really like what you're doing and you're good at it, you will be successful." Nowadays I get paid to have fun doing things I'd do for free anyway, so that meets my definition of success! Thanks, Dad.

Acknowledgment

This book is a direct result of contributions from many of the students I have been fortunate enough to have in my embedded computer engineering courses at the University of California—San Diego extension. They have provided a valuable form of feedback by sharing their notes and pointing out weaknesses in the text and in-class presentations. Some sections of this text were also provided by David Fern and Steven Tietsworth.

I would also like to thank my family for supporting me and, Mary, Nikki, Kenny, Daniel, Amy, and Annie for being patient and helping out when I needed it!

A Volume in the
Embedded Technology™ Series

Embedded Controller Hardware Design

by Ken Arnold

Technology
Publishing

www.LLH-Publishing.com
www.EmbeddedControllerHardwareDesign.com

ii

ISBN: 1-878707-52-3

Library of Congress Control Number: 00-135391

Printed in the United States of America

10 9 8 7 6 5 4 3 2 1

Project management and developmental editing: Harry Helms, LLH Technology Publishing

Interior design and production services: Greg Calvert, Model, CO

Cover design: Sergio Villareal, Vista, CA

LLH Technology
Publishing

www.LLH-Publishing.com
www.EmbeddedControllerHardwareDesign.com

Table of Contents

Preface

During the early years of microprocessors, there were few engineers with education and experience in the applications of microprocessor technology. Now that microprocessors and microcontrollers have become pervasive in so many devices, the ability to use them has become almost a requirement for many technical people.

Today the microprocessor and the microcontroller have become two of the most powerful tools available to the scientist and engineer. Microcontrollers have been embedded in so many products that it is easy to overlook the fact that they greatly outnumber personal computers. Millions of PCs are shipped each year, but *billions* of microcontrollers ship annually. While a great deal of attention is given to personal computers, the vast majority of new designs are for embedded applications. For every PC designer, there are thousands of designers using microcontrollers in embedded applications. The number of embedded designs is growing quickly. The purpose of this book is to give the reader the basic design and analysis skills to design reliable microcontroller or microprocessor based systems. The emphasis in this book is on the practical aspects of interfacing the processor to memory and I/O devices, and the basics of interfacing such a device to the outside world.

A major goal of this book is to show how to make devices that are inherently reliable by design. While a lot of attention has been given to "quality improvement," the majority of the emphasis has been placed on the processes that occur *after* the design of a product is complete. Design deficiencies are a significant problem, and can be exceedingly difficult to identify in the field. These types of quality problems can be addressed in the design phase with relatively little effort, and with far less expense than will be incurred later in the process. Unfortunately, there are many hardware designers and organizations that, for various reasons, do not understand the significance and expense of an unreliable design. The design methodology presented in this text is intended to address this problem.

Learning to design and develop a microcontroller system without any practical hands-on experience is a bit like trying to learn to ride a bike from reading book. Thus, another goal is to provide a practical example of a complete working product. What appears easy on paper may prove extremely difficult without some real world experience and some potentially painful crashes. In order to do it right, it's best to examine and use a real design. On the other hand, the current state of the technology (surface mounted packaging, etc.) can make the practical side problematic. In order to address this problem, a special educational System Development Kit is available to accompany this book (8031SDK). All the documentation to construct an SDK is available on the companion CD-ROM. This info, along with updated information and application examples, is also available on the web site for this book: http://www.hte.com/echdbook. All the information needed to build the SDK is available there, as well as information on how to order the SDK assembled and tested.

While searching for an appropriate text for one of the courses I teach in embedded computer engineering, I was unable to locate a book that covered the topic adequately. An earlier version of this book was written to accompany that course and has since evolved into what you see here. The course is offered at the University of California, San Diego Extended Studies, and is titled "Embedded Controller Hardware Design." The same courses may also be taken in an on-line format using the Internet, and can be found at http://www.hte.com/uconline/ecd The goals of the course and the book are very much the same: to describe the *right way* to design embedded systems.

While no prior knowledge of microcontrollers or microprocessors is required, the reader should already be familiar with basic electronics, logic, and basic computer organization. Chapter one is intended as a review of those basic concepts. Next there is a general overview of microcontroller architecture, and a specific microcontroller chip architecture, the 8051 family, is introduced and detailed. The 8051 was chosen because it can be interfaced to external memory, has simple timing specs, is widely used and available from a number of manufacturers. The concepts of worst-case design and analysis are described, along with techniques for hardware interfacing. A good embedded design requires familiarity with the underlying memory technology, including ROM, SRAM, EPROM, Flash EPROM, EEPROM storage mechanisms and devices. The processor bus interface is then covered in general form, along with an introduction to the 8051's bus interface. Most embedded designs can also benefit from the use of user programmable logic devices (PLD). This subject is too complex for in-depth coverage here, so PLD technology is covered from a relatively high level. The central theme of designing an embedded system that can be proven to be reliable is illustrated with a simple embedded controller. The iterative nature of the design process is shown by example, and several design alternatives are evaluated. With the central part of the design completed, the remaining chapters cover the various types of I/O interfaces, bus operations, and a collection of information that is seldom included in the usual sources, but is often handed down from one engineer to another.

I hope that you will find this book to be useful, and welcome any observations and contributions you may have. If you should find any errors in the text, or if you know of some good embedded design resources, please feel free to contact me directly by e-mail: ken.arnold@ieee.org

Review of Electronics Fundamentals

Why are microprocessors and microcontrollers designed into so many different devices? While there are many dry and practical reasons, I suspect one of the strongest motivations for using a microprocessor is simply that it is a lot more fun.

Over the past few decades of the so-called "computer revolution," I have seen many products and projects that could have been handled without resorting to a microprocessor. Yet there is always a tendency to rationalize the choice of a micro-based solution by economic or technical arguments to support the decision. In fact, most of the really excellent products were successful to a great extent because they were fun to develop. Many of the best product ideas have occurred when someone was "playing" with something they were interested in. In my own experience, I have found learning something new is much easier and more effective when I am "just playing around" rather than trying to learn in a structured way or against a deadline. Studies of various educational methods also indicate "coached exploration" is more effective than the traditional methods. These and other observations lead me to the conclusion that the best way to learn about a microcontroller is by "playing" with one.

No book—no matter how well written—can possibly motivate and educate you as well as building and playing with a microcontroller. The best way to learn the concepts in this book is to build a simple microcontroller. Even if it is capable of nothing more than blinking a light, it will provide a concrete example of the microcontroller as a tool that can be fun to use. To ease this effort, a companion system development kit (SDK), is available to accompany this text. It incorporates the functions of a stand-alone single board computer (SBC), and an in-circuit emulator (ICE). It also serves as a sample embedded controller design. The design is included on the CD-ROM and web site for this book, so anyone can reproduce and use it as a learning tool. By applying the guidelines set forth in this book to real world hardware, you can learn to design reliable embedded hardware into other products. Information on obtaining the SDK can be found in the Preface.

Objectives

Several different skills are required for successful embedded hardware design. Here are some of the things you will know how to do when you finish this book:

- Interpret design requirements for the design of an embedded controller.
- Read and understand the manufacturer's specification sheet.
- Select appropriate ICs for the design.
- Interface the CPU, memory, and I/O devices to a common bus.
- Design simple I/O (input/output) interfaces.
- Define the decoding and interconnection of the major components.
- Perform a worst-case analysis of the timing and loading of all signals.
- Understand the software development cycle for a microcontroller.
- Debug and test the hardware and software designs.

These tasks represent the major skills required in the successful application of an embedded micro. In addition, other abilities—such as the design and implementation of simple user programmable logic—will be covered as required to support the proficient application of the technology.

Embedded Microcomputer Applications

There is an incredible diversity of applications for embedded processors. Most people are aware of the highly visible applications, but there are many less apparent uses. Many of the projects my students have chosen turned out to be of practical use in their work. However, they have covered the entire range from the economically practical to the blatantly absurd. One practical example was the use of a microprocessor to monitor and control the ratio of ingredients used in mixing concrete. About a year after the student implemented the system, he wrote to inform me that the system had saved his company between two and three million dollars a year by reducing the number of "bad batches" of concrete that had to be jack hammered out and replaced. Another example was that of a student who suspended a ball by airflow generated by a fan and provided closed loop control of the ball's position with the microprocessor. The only thing that many of the student projects really had in common was the use of a microcontroller as a tool.

Some of the actual commercial applications of embedded computer controls that the author has been directly involved with include:

- A belt measures a person's heart rate and respiration that signals an alarm when safe limits are exceeded. A radio signal is then transmitted to a microcontroller in a pocket pager to display the type of problem and the identity of the belt.
- An environmental system controls the heating ventilating and air conditioning in one or more large buildings to minimize peak energy demands.
- A system that measures and controls the process of etching away the unwanted portions of material from the surface of an integrated circuit being manufactured.
- The fare collection system used to monitor and control entry to a rapid transit system based on the account balance stored on the magnetic stripe on a card.
- Determination of exact geographic position on the earth by measuring the time of arrival of radio signals received from navigational beacons.
- An intelligent phone that receives radio signals from smoke alarms, intrusion sensors, and panic switches to alert a central monitoring station to potential emergency situations.
- A fuel control system that monitors and controls the flow of fuel to a turbine jet engine.

Selecting a particular processor for a given application is usually a function of the designer's familiarity with a particular architecture. While there are many variations in the details and specific features, there are two general categories of devices: microprocessors and microcontrollers. The key difference between a *microprocessor* and a *microcontroller* is that a microprocessor contains only a central processing unit (CPU) while a microcontroller has memory and I/O on the chip in addition to a CPU. Microcontrollers are generally used for dedicated tasks. *Microcomputer* is a general term that applies to complete computer systems implemented with either a microprocessor or microcontroller.

Microcomputer and Microcontroller Architectures

Microprocessors are generally utilized for relatively high performance applications where cost and size are not critical selection criteria. Because microprocessor chips have their entire function dedicated to the CPU and thus have room for more circuitry to increase execution speed, they can achieve very high-levels of processing power. However, microprocessors require external memory and I/O hardware. Microprocessor chips are used in desktop PCs and workstations where software compatibility, performance, generality, and flexibility are important.

By contrast, microcontroller chips are usually designed to minimize the total chip count and cost by incorporating memory and I/O on the chip. They are often "application specialized" at the expense of flexibility. In some cases, the microcontroller has enough resources on-chip that it is the only IC required for a product. Examples of a single-chip application include the key fob used to arm a security system, a toaster, or hand-held games. The hardware interfaces of both devices have much in common, and those of the microcontrollers are generally a simplified subset of the microprocessor. The primary design goals for each type of chip can be summarized this way:

• microprocessors are most flexible

• microcontrollers are most compact

There are also differences in the basic CPU architectures used, and these tend to reflect the application. Microprocessor based machines usually have a *von Neumann architecture* with a single memory for both programs and data to allow maximum flexibility in allocation of memory. Microcontroller chips, on the other hand, frequently embody the *Harvard architecture*, which has separate memories for programs and data. Figure 1-1 illustrates this difference.

Figure 1-1: At left is the von Neumann architecture; at right is the Harvard architecture.

One advantage the Harvard architecture has for embedded applications is due to the two types of memory used in embedded systems. A fixed program and constants can be stored in non-volatile ROM memory while working variable data storage can reside in volatile RAM. Volatile memory loses its contents when power is removed, but non-volatile ROM memory always maintains its contents even after power is removed.

The Harvard architecture also has the potential advantage of a separate interface allowing twice the memory transfer rate by allowing instruction fetches to occur in parallel with data transfers. Unfortunately, in most Harvard architecture machines, the memory is connected to the CPU using a bus that limits the parallelism to a single bus. A typical embedded computer consists of the CPU, memory, and I/O. They are most often connected by means of a shared bus for communication, as shown in Figure 1-2.

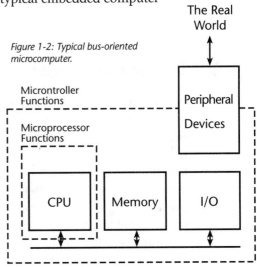

Figure 1-2: Typical bus-oriented microcomputer.

The peripherals on a microcontroller chip are typically timers, counters, serial or parallel data ports, and analog-to-digital and digital-to-analog converters that are integrated directly on the chip. The performance of these peripherals is generally less than that of dedicated peripheral chips, which are frequently used with microprocessor chips. However, having the bus connections, CPU, memory, and I/O functions on one chip has several advantages:

- Fewer chips are required since most functions are already present on the processor chip.
- Lower cost and smaller size result from a simpler design.
- Lower power requirements because on-chip power requirements are much smaller than external loads.
- Fewer external connections are required because most are made on-chip, and most of the chip connections can be used for I/O.
- More pins on the chip are available for user I/O since they aren't needed for the bus.
- Overall reliability is higher since there are fewer components and interconnections.

Of course there are disadvantages too, including:

- Reduced flexibility since you can't easily change the functions designed into the chip.

- Expansion of memory or I/O is limited or impossible.

- Limited data transfer rates due to practical size and speed limits for a single-chip.

- Lower performance I/O because of design compromises to fit everything on one chip.

Digital Hardware Concepts

In addition to the CPU, memory, and I/O building blocks, other logic circuits may also be required. Such logic circuits are frequently referred to as *glue logic* because they are used to connect the various building blocks together. The most difficult and important task the hardware designer faces is the proper selection and specification of this "glue logic." Devices such as registers, buffers, drivers and decoders are frequently used to adapt the control signals provided by the CPU to those of the other devices. While TTL gate level logic is still in use for this purpose, the *programmable logic device* (PLD) has become an important device in connecting the building blocks. Contemporary microcontroller designers need to acquire the following skills:

- Interpretation of manufacturers specifications
- Detailed, worst case timing analysis and design
- Worst case signal loading analysis
- Design of appropriate signal and level conversion circuits
- Component evaluation and selection
- Programmable logic device selection and design

The glue logic used to join the processor, memories, and I/O is ultimately composed of logic gates, which are themselves composed almost entirely of transistors, diodes, resistors, and interconnecting wires. In order to understand the basic operation of the glue logic, we are going to begin at the component level with a review of basic electronics concepts. These concepts will be presented as fluid flow analogies.

Voltage, Current, and Resistance

In Figure 1-3, a battery provides a voltage source for electricity, much like a pump provides a pressure source for a fluid. Voltage, or pressure, is required to produce current flow in the circuit.

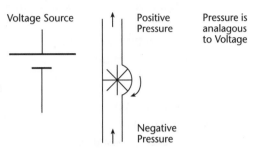

Figure 1-3: Voltage in an electrical circuit is analogous to pressure in a fluid.

The voltage source provides the pressure "motivation," if you will, for current flow. Resistance provides a limiting constraint on the amount of current that will actually flow. The resistor will allow a current to flow through it that is proportional to the voltage across it, and inversely proportional to the resistance value. Higher resistance is like a smaller aperture for the fluid to flow through. The resistance results in a voltage, or pressure drop, across the resistance as long as current is flowing in the resistor. Figure 1-4 illustrates this.

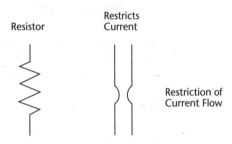

Figure 1-4: Resistance in an electrical circuit is analogous to a restriction in the flow of a fluid.

The wiring connecting the components in a circuit is like the piping connecting plumbing components that let a fluid flow. The flow of current in the circuit is controlled by the magnitude of the voltage (pressure) and the resistance (pressure drop) in the circuit. In Figure 1-5, the battery provides a voltage to force current through the resistor. The magnitude of the voltage (V) generated by the battery is developed across the resistor, and the magnitude of the resistance (R), determine the current (I). Note the "return" current path is often shown as "ground," which is the reference voltage used as the "zero volts" point. In this case, current flows from the positive battery terminal, through the wire, then the resistor, then through the "ground" connection to the minus terminal of the battery. This is usually not the same as earth ground, which provides a connection to a stake or pipe literally stuck in the ground. The magnitude of the current in this case is $I = V / R$ by re-arranging the equation V = I * R, as shown in Figure 1-5. This is known as Ohm's law. Another way to look at it is that whenever current flows through a resistor, there is a drop in voltage across the resistor due to the restriction in current.

Real components are not the perfect voltage sources, resistances, etc. we have discussed so far. They have parasitic values that limit

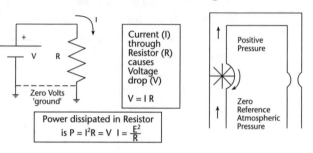

Figure 1-5: Voltage across R is equal to current multiplied by resistance.

their performance in the real world and are subject to other limitations, such as operating temperature, power limits, etc. Current flows only through a complete circuit, and in most cases (for a positive power supply) current flows from the power source through the circuitry and returns to the power supply through the common "ground" connection. Current flowing through any resistance results in the dissipation of power as heat. The power dissipated is P = I²R = V*I = V²/R. Note that voltage is sometimes denoted by the variable V and sometimes by E, for *electromotive force*.

All practical components have some resistance. Real batteries have an internal resistance, for example, which provides an upper limit to the current the battery can supply to an external circuit. Real wires have resistance as well, so the actual performance of a circuit will deviate somewhat from the ideal. These effects are obvious in some cases, but not in others. In an automobile starting circuit, it's not surprising that the battery, supplying 12 volts to a starter with internal resistance on the order of 0.01 to 0.1 ohms, will result in currents of hundreds of amperes in order to start the engine. On the other hand, while consulting with a prominent notebook computer manufacturer, I uncovered a design error resulting in an internal current of hundreds of amperes flowing in the circuit for a few nanoseconds. Obviously, this wreaked havoc on the operation of the computer, and generated a great deal of electromagnetic noise!

One of the things you will learn in this book is how to avoid those kinds of mistakes. It's also important to remember that power is dissipated in *any* resistance present in the circuit. The power is proportional to the voltage times the current across the resistance, which is dissipating the power. In the last two examples, the amount of power dissipated instantaneously is quite high while the current is flowing. When the current pulse is only a few nanoseconds long, however, it may not be obvious, since there won't be much heat generated.

Diodes

The diode is a simple semiconductor device acting as a "one way" current valve. It only lets current flow in one direction. Figure 1-6 illustrates how the diode operates like a "one-way" fluid valve.

(*Purists please note:* This book does not use electron current flow. All electrical current flow will be "positive" or "conventional" current flow, meaning current

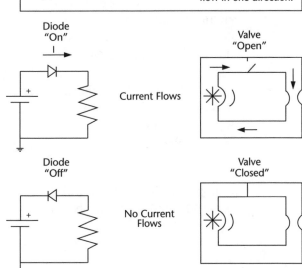

Figure 1-6: A diode to electricity is analogous to a valve in the flow of a fluid.

always flows from the most positive terminal to the most negative terminal of a component. The use of positive current flow follows the intuitive direction of the arrows inherent in the component drawings for diodes, transistors, etc.)

Transistors

The flow analogy can also be used to model how a transistor operates in a logic circuit. The transistor is an amplifier. It uses a small amount of energy to control a larger energy source, just as a valve controls a high-pressure water source. There are two kinds of transistors: *bipolar* and *field-effect transistors* (FETs). We will look at bipolar transistors first; these amplify current. A small amount of current flows in the control circuit (the transistor base-emitter circuit) to turn the transistor on. This control current is amplified (multiplied by the gain or *beta* of the transistor) and allows a larger current to flow in the output circuit (the collector-emitter circuit). Once again, the device is not perfect because of the resistance, current, gain, and

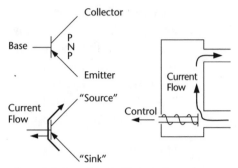

Figure 1-7: Operation of a bipolar PNP transistor.

leakage limitations of real transistors. Bipolar transistors come in two polarities, NPN and PNP, with the difference being the direction in which current flows for normal operation. A bipolar PNP transistor is shown and modeled in Figure 1-7.

For most of the illustrative circuit examples in this book, we will be using NPN transistors, as shown in Figure 1-8.

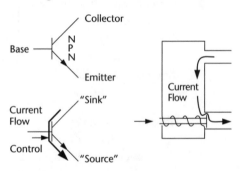

Mechanical Switches

Figure 1-8: Operation of a bipolar NPN transistor.

Mechanical switches are useful for direct input to digital circuits. One of the more convenient versions is a bank of rocker switches packaged into a module that can fit into the same location as a standard chip. The *dual in-line package*, or DIP, switch is one of the easiest ways to add multiple switches to a microcontroller design. The mechanical switch has extremely low "on" resistance and high "off" resistance, unlike most semiconductor switches. Figure 1-9 shows a typical DIP switch and the schematic symbol for it.

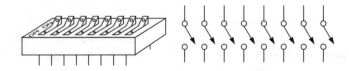

Figure 1-9: 8-position DIP switch and schematic equivalent.

Transistor Switch ON

Transistors can be configured to function as switches. As can be seen in
Figure 1-10, an NPN transistor operating as a current controlled switch can
be used to build a simple inverter. It changes a logic one on its input to a logic
zero at its output, and vice versa. In this case, logic one is represented as a
positive voltage, and a logic zero is represented by zero volts. The logic one
input (positive input voltage) is supplied through a resistor from the power
supply voltage to the transistor base terminal, resulting in a small base control
current into the base.

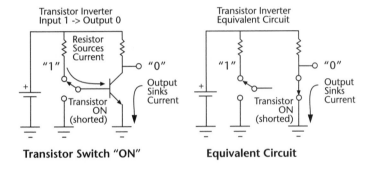

Transistor Switch "ON"   **Equivalent Circuit**

*Figure 1-10: The transistor inverter; input = 1 and transistor ON. The transistor
ON configuration is at left and the equivalent circuit is at right.*

The transistor is used because it has gain allowing a larger output current
to flow as controlled by a weaker input. When the transistor is turned on
as much as it can be, the collector emitter circuit looks almost like a short
circuit, effectively connecting the output to ground or zero volts. This gives
a logic zero on the collector output. When the transistor collector is shorted
to ground, current flows from the supply through the resistor and into the
transistor collector to ground. The transistor is said to *sink* the resistor
current into ground. If there is an external load, such as another inverter or
gate, connected to the collector output, the transistor can also sink current
from the load. This is also referred to as *pulling down* the output voltage.
The current sinking capacity of the transistor limits the number of devices
this inverter can drive.

Transistor Switch OFF

When the input is connected to logic zero (ground voltage), no current flows into the base of the transistor, since its base and emitter terminals are at the same voltage. When there is no current flowing in the base, the transistor will not allow current to flow in the collector emitter circuit either. As a result, the circuit behaves as if the transistor was removed from the circuit. The output resistor will source current to any potential load. The output is pulled up to the supply voltage, resulting in a logic one at the output. Once again, there is a limit to the resistor's ability to source current, resulting in a limit to the number of loads that can be attached to this circuit's output. Notice these two limits are defined by the ability of the transistor to pull down the output, and the resistor's ability to pull up the output become the main limits to its ability to drive other devices. Gates can be constructed by adding diodes or transistors to the inverter circuit in Figure 1-11.

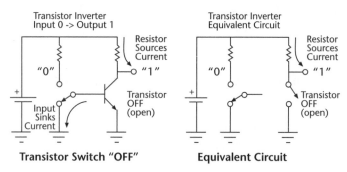

Figure 1-11: The transistor inverter; input = 0 and transistor OFF.
The transistor OFF configuration is at left and the equivalent circuit is at right.

The FET as a Logic Switch

Most of the logic devices used in highly integrated circuits do not use bipolar transistors. Instead, they use field effect transistors. FETs perform a similar function to the bipolar transistors discussed earlier, but they are voltage controlled. While the current flowing in the base controls bipolar transistors, the voltage between the gate and source controls field effect transistors. The gate voltage of a field effect transistor controls the current flowing in the drain-source circuit. The symbol for the FET shows the gate to be insulated from the source-drain circuit, as shown in Figure 1-12.

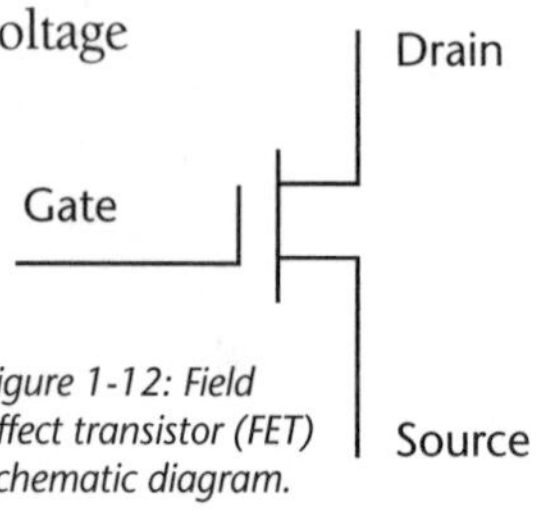

Figure 1-12: Field effect transistor (FET) schematic diagram.

This type of FET is referred to as a MOSFET (*metal oxide semiconductor* FET), since the insulating material is silicon dioxide (SiO_2), commonly known as glass (for early devices, the gate was made of metal). Like bipolar NPN and PNP transistors with opposite polarity, FETs come in N- and P- channel varieties. The N- and P- channels refer to the polarity of the source drain element of the device. A cross-section view of a FET is shown in Figure 1-13.

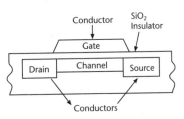

Figure 1-13: Field effect transistor cross-section.

NMOS Logic

The conductive state of the FET's channel is what allows or prevents current from flowing in the device. For a typical logic N-channel MOSFET, the channel becomes conductive when the gate has a positive voltage with respect to the source, allowing current to flow between the drain and source terminals. When the gate is at the same voltage as the source, no current flows. The design of MOS logic circuits can be almost exactly equivalent to the bipolar inverter we saw earlier, substituting an N-channel MOSFET for the bipolar NPN transistor. In fact, the most of the early microcontroller integrated circuits were manufactured using variations of this method, and are referred to as *NMOS* logic. As can be seen from Figure 1-14, the NMOS FET circuit behaves in an equivalent way to the NPN transistor inverter. When the gate (control input) of the NMOS FET is at a positive voltage, the FET is ON, effectively shorting the source and drain pins. When the gate is at 0 volts, the FET is OFF, opening the circuit between the source and drain. Older NMOS logic ICs use this type of circuit. The original 8051 microcontroller was an NMOS processor.

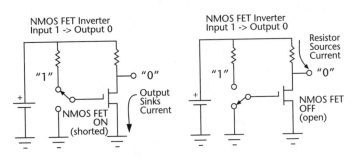

Figure 1-14: NMOS inverter circuit.

CMOS Logic

CMOS logic (*complementary* symmetry MOS) is another form of MOS logic. It has advantages over NMOS logic for low power circuitry and for very complex integrated circuits. NMOS logic is relatively simple, but it has one serious drawback: it consumes a significant amount of power. In fact, it would be impossible to manufacture the largest ICs using NMOS logic, as the power dissipated by the chip would cause it to overheat. This is the main reason CMOS logic has become the dominant form of logic used for large, complex ICs. Instead of using a resistor to source current when the output is high, a CMOS device uses a P-channel MOSFET to pull the output high. CMOS logic is based on the use of two complementary FETs that switch the output between the power supply and ground. A simple CMOS inverter is shown in Figure 1-15.

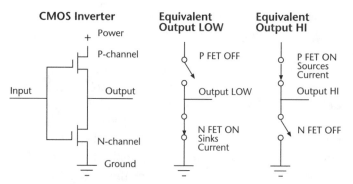

Figure 1-15: CMOS inverter circuit and equivalent output.

CMOS logic uses two switches: one P-channel pull-up transistor, and one N-channel pull-down device to pull the output low or high, one at a time. CMOS logic is designed with an N-channel device that turns on and conducts when the gate voltage is at logic one (positive voltage), and the P-channel device turns on when the gate is at ground voltage. A CMOS inverter is comprised of a pair of FETs, one device of each type, as shown in Figure 1-15.

When the transistor gate inputs are at logic one (positive voltage), the P-channel device is off, and the N-channel device is on, effectively connecting the output to ground, or logic zero. Likewise, when the input is grounded, the P-channel device turns on and the N-channel device turns off, effectively connecting the output to the positive supply voltage, or logic one. Gates and more complex logic functions can be constructed by using series and parallel-connected MOSFETs in circuits similar to the one above. The gate of a

MOSFET, as implied by the symbol, is essentially an open circuit. In fact, the gate of a MOSFET does have an *extremely* high resistance. The operation of the MOSFET's channel is controlled by the voltage of the gate, unlike the bipolar NPN transistor we examined in the inverter, which is controlled by input (base) current. Bipolar transistors are current amplifiers, with their output current being controlled by their base current. FET outputs, on the other hand, are dependent on the gate voltage.

Since almost no current flows in a CMOS output when it is driving a CMOS gate input in the steady state condition, these logic devices consume much less power than the other types. MOS logic has some other advantages over bipolar logic, since there is almost no input current (less than one nanoampere, or 10^{-9} ampere), so it does not need to exact a DC current load on the device driving it. This is good news, because it means that the input current of a CMOS device does not limit the number of gates that can be connected to the output of the driving gate. The number of gate inputs that a single gate output can drive is the gate *fan-out*. Fan-out applies between gates of the same logic family, as different families of logic have different output capabilities and their inputs present different loads.

Now for the bad news about the high input resistance of MOS devices: the insulation separating the input from the channel is very thin (measured in angstroms). This thin layer can easily be punctured by electrostatic discharge (ESD), such as occurs regularly when dissimilar materials rub against one another. Just walking across the room can generate tens of kilovolts, which is more than enough to destroy a MOS device. As a result, special precautions must be taken to prevent damage to MOS devices. When handling these devices, it is important to ground your body *before* touching the device, and to also keep the device at or near ground. Special wrist straps and workspace mats are available to assist in keeping static voltages from building up and for dissipating them when they do occur. Special, conductive bags and containers should be used when possible to contain sensitive devices.

CMOS power consumption is usually dominated by the power consumed during the transition of a logic device from one state to another. As a result, pure CMOS devices consume only a few microamperes of current when they are not switching, and the bulk of the current drawn is a function of clock frequency. The higher the clock frequency, the greater the current consumption. For pure CMOS, the power supply current is linearly proportional to the clock rate.

Mixed MOS

Many logic devices labeled as CMOS are actually a mixture of NMOS and CMOS, because the manufacturer needs to compromise the extremely low power of CMOS with the performance of NMOS logic. This can be a problem for design-ers of battery-powered systems, since the current requirement (and the resulting battery life) of a pure CMOS circuit is orders of magnitude better than an NMOS circuit. Many CMOS memories are actually mixed MOS, and are not appropri-ate for battery-powered systems. True CMOS chips can retain their contents for years using only a single coin cell to maintain power to the memory.

Real Transistors Don't Eat Q!

So far we have described the various types of transistors as perfect switches that have zero resistance when they're on and infinite resistance when they're off. When we examine the actual behavior, we find that real transistors do not exhibit these characteristics. A transistor switch may have tens or hundreds of ohms of resistance when it is on, and hundreds or even tens of thousands of ohms of "leakage" resistance when it's off. As a result, the logic outputs aren't perfect either. When the tran-sistor is on, the output voltage is a function of the output current, due to the voltage drop across the resistance. As Figure 1-16 shows, the output voltage of a logic device will depend upon how much cur-rent is flowing in the output and the resistance of the switch.

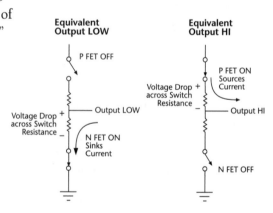

Figure 1-16: Logic output voltage is current dependent.

Unfortunately, the switch resistance is also non-linear so that the switch resis-tance changes as the voltage across the switch changes. This makes it difficult to picture the output behavior under different operating conditions. The behavior will also differ from one device to another, over temperature, and so on. Manufacturers only specify the output characteristic at one point on the

curve, Vo at Io max. As a result,
the best we can do is to look
at the output characteristics
graphically, as shown in
Figure 1-17.

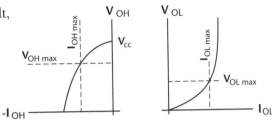

Figure 1-17: Output voltage Vo versus current Io.

Logic Symbols

Logic symbols are used to represent the logic functions in a more abstract way,
allowing the designer to specify the logical function of a circuit without getting
into the details of the underlying components (such as the transistors and
resistors). The logic symbols used in this text represent those that are most
commonly used in commercial documentation. There are other standards, such
as the ANSI/IEEE standard gate level symbols, but they are not encountered
as frequently in practice. Figure
1-18 shows the logic symbols for
different gates, and their functions
are described in the truth tables.

The logic symbols in Figure 1-18
show the shapes and Boolean logic
functions for the most common
gate configurations. The buffer
device is a triangle—the symbol
for an amplifier—because it
amplifies the input signal, allowing
an increase in the number of loads
that can be driven. Note that a
small circle, often referred to as a
"bubble," on an input or output
terminal designates a logical inversion. Thus the inverter is shown as

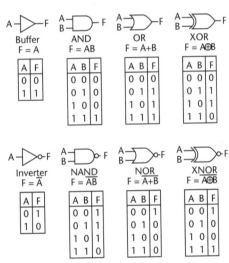

Figure 1-18: Logic symbols, symbolic notation,
and truth tables.

a triangle (amplifier) with a bubble on the output to signify the logic level
inversion on the output. The logic voltage levels for TTL logic are:

Positive Logic
0 = false = lowest voltage level
1 = true = highest voltage level

Corresponding TTL Logic Voltages
0 = input voltages 0 to 0.8 volts (low)
1 = input voltages 2 to 5 volts (high)

This means that a TTL compatible logic input is guaranteed to respond to an input signal between 0 and 0.8 volts as a logic zero, and input voltages from 2 to 5 volts as a logic one. Note that voltages between 0.8 and 2 volts are *not* valid logic levels.

Logic voltage levels are different for different types of logic, but the most common logic levels are those corresponding to the original TTL (*transistor-transistor logic*), using a 5 volt power supply. CMOS levels, using 3 or 5 volt power, are also common. TTL and CMOS logic—like almost every other type of logic in common use —are called *positive logic* because the most positive voltage corresponds to the logic one value.

Tri-State Logic

Tri-state logic does not refer to orderly thinking in a three state geographic region! When we speak of binary (base two number) values, we mean that a given bit or logic signal can take on either one of two valid states (zero or one) at any instant in time. A logic gate that is not forcing its output to be either one or zero is said to be *tri-stated*. Tri-state logic does not refer to base three numbers, but rather to a third invalid logic state when the output of a logic device is neither sinking nor sourcing current. This so-called third state is really an undefined condition, because the device output is not forcing a logic level on its output. It is said to be in a floating, high imped-ance, passive, or Hi-Z state, since the output circuits are effectively disconnected. A tri-state driver connected to one signal wire of the bus is shown in Figure 1-19.

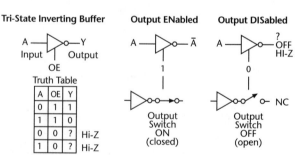

Figure 1-19: Active and passive states of a tri-state buffer.

On the left is an inverting buffer with an enabled tri-state output. On the right side is an example showing two of the same type of buffers, with the top device in the disabled or passive state, and the lower device is enabled or actively driving the data bus to a logic one level. The control signal determines whether the output is passive or active, and is called the output enable or OE signal. The device shown above is actively driving the bus whenever the OE control line is at a logic one level, and is passive when the OE line is at a logic zero level. Most of the time, output enable signals are *active low*, meaning that the output is enabled when the /OE signal is low, and passive when the /OE signal is high. This is shown on the logic symbol with an inversion bubble where the enable signal enters the logic device.

As computer circuits become more dense and complex, the connecting wires have become increasingly difficult to route and interconnect. This is especially true on a densely packed integrated circuit, where it turns out that the wiring is more valuable than the logic gates! On one common CPU chip, 68% of the chip area is used for interconnect wiring. Even on a circuit board, it is important to use the board wiring in an efficient way. Since there are many parallel address and data lines that must go to multiple chips, the multiplexing approach makes it practical to connect many devices. The purpose for using tri-state logic is to allow multiple devices to share wires by taking turns one at a time. This may sound a bit silly, but it is just one form of multiplexing, or sharing a resource that needs to be allocated among multiple devices. When the resource is a collection of parallel data wires, referred to as a *data bus*, and the bus is shared by multiple microcomputer CPU and peripheral devices transferring information one at a time in sequence, it is referred to as a *multiplexed data bus*.

Timing Diagrams

The timing diagram is the standard "language" of illustrating timing relationships between different parts of a design. In order to understand the relationship of different signals with respect to time, it is necessary to learn how to read and interpret timing diagrams. Figure 1-20 shows examples of *asynchronous* (un-clocked or combinatorial gates) and *synchronous* (clocked flip-flop) logic. The notation used in this book is representative of that used in most component specifications. Timing specifications, such as delay, setup, and hold times, specify the limits under which the device is guaranteed to operate as intended. If those specifications are violated, the device may very well operate correctly most of the time. However, a change in temperature, voltage, or variations from unit to unit may make the circuit unreliable. The most undesirable result of timing violations is that the circuit makes very infrequent errors, perhaps one error in hundreds of hours of operation. If you have ever wondered why your PC crashes mysteriously for no apparent reason, timing specification violations may well be the cause!

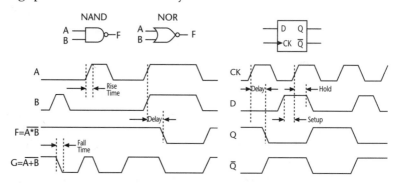

Figure 1-20: Timing diagram notation examples.

Timing relationships are particularly important for signals that are "time shared" on a single wire. A group of wires that carries different information at different times is also called a bus.

Multiplexed Bus

In order to describe the timing of such a shared data bus, it is necessary to define some notation for timing diagrams. The notation used in this book is shown in Figure 1-21.

The terminology for timing parameters is covered in a later chapter, but the basic concept for time multiplexed data on a bus is shown in Figure 1-21. The two devices are alternately enabled to drive the data bus wire, allowing each to drive the bus in turn. Only one device is allowed to drive the bus at a time when it is operating correctly.

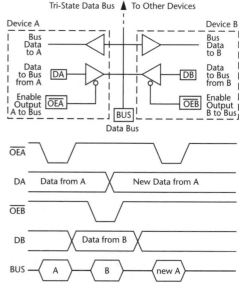

Figure 1-21: Time multiplexed data bus and timing.

Timing diagrams are a critical method to allow accurate and unambiguous representation of the time related operations of digital circuits, which we will be using to understand and document the correct sequence of operations for microcomputer systems. Timing analysis, using these diagrams, allows the designer to determine safe and reliable limits to proper operation of the various circuits in the system. It is better to take a little more time to design a circuit correctly from the start than it is to find and fix bugs during testing. This is especially true because of the increasing cost of fixing a bug as a product progresses through production and into the field.

Loading and Noise Margin Analysis

In addition to timing, the designer must consider the voltages and loads at the logic inputs and outputs. If the output of one gate is connected to the input of another, the designer must assure that the logic voltages are compatible. Once again, just as for the timing, violations of these specifications often result in infrequent errors that are very tricky to reproduce. Again, prevention is much simpler than tracking down bugs as they appear in production units. This topic is the subject of Chapter Three.

The Design and Development Process

Structured design of a microcomputer requires the ability to do the system design and partitioning from the top down while implementing the system from the bottom up. The hardware design and development process should consist of the following steps:

1) Defining the requirements.
2) Collecting information on potential components.
3) Evaluate the components with respect to the requirements.
4) Do a block diagram preliminary design and component selection.
5) Perform a preliminary timing and loading analysis.
6) Define the functions of the "glue logic."
7) Schematic entry using CAD (computer-aided design) software.
8) Programmable logic device design and simulation.

9) Detailed timing analysis and simulation, adjusting the design as required.

10) Check the signal loading, buffering signals as needed.

11) Document the design and generate a net list and bill of materials.

12) Begin the design and layout of a printed circuit board.

13) Implement the design in breadboard or prototype form.

14) Program the memories and programmable logic as required for testing.

15) Debug and verify operation using oscilloscope, logic analyzer, and in-circuit emulator.

16) Update and complete documentation as the design changes.

The order of tasks shown is variable, and some of the tasks may be performed in parallel. Software design is also frequently done in parallel with hardware design, and sometimes even before the hardware design. This is frequently a result of the fact that the cost and time required to develop the software exceeds that of the hardware development. In some cases the cost of modifying existing programs may be so high as to be impractical. In these cases, it is the designer's responsibility to maintain software compatibility with previous hardware designs.

Chapter One Problems

1. If an open-drain N-channel FET transistor is used as a logic output, is it possible to connect more than one open-drain transistor output to the same signal? What would the effect of doing so be on the resulting combined signal?

2. If a logic output sinks I_{OL} = 10 milliamperes with an output voltage, V_{OL} = 0.5 volts, how much power is dissipated by a 450 ohm resistor between the output and the 5 volt power supply?

3. How much current must a logic output source, in order to maintain an output voltage of 2.5 volt when driving a 5 kilohm resistor connected to ground?

4. In a CMOS inverter, there is a short period of time when both the N- and P-channel transistors are partially turned on when the input is changing from low to high or high to low. What effect will this have on power consumption? What characteristic in the input signal would reduce this effect?

Microcontroller Concepts

One way of looking at a computer system is to consider the successive "translations" that occur from the high level code (a programming language such as C++) to the electrical signals that "communicate" with the hardware. A computer system can be broken down into multiple levels or layers to show the translation of a specific instruction into a form that can be directly processed by the computer hardware. Such hierarchical levels are discussed in detail in *Structured Computer Organization* by A.S. Tanenbaum. This hierarchy is shown in Figure 2-1

High Level   Sum := Sum + 1

Assembly   MOV BX,SUM   INC   (BX)

Machine   1101010100001100 0010001101110101 1111100011001101

Register Transfer   Fetch Instruction, Increment PC, Load ALU with SUM ...

Gate

Circuit

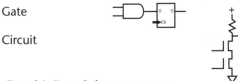

Figure 2-1: "Layers" of a computer system.

Language translators such as compilers and assemblers translate high-level code into machine code that can be executed by the processor. The primary focus of this book will be from the assembly and machine language level downward.

Organization: von Neumann vs. Harvard

We introduced the von Neumann and Harvard computer architectures in Chapter One. The von Neumann machine, with only one memory, requires all instruction and data transfers to occur on the same interface. This is sometimes referred to as the "von Neumann bottleneck." In common computer architectures, this is the primary upper limit to processor throughput. The Harvard architecture has the potential advantage of a separate interface allowing twice the memory transfer rate by allowing instruction fetches to occur in parallel with data transfers. Unfortunately, in most Harvard architecture machines, the memory is connected to the CPU using a bus that limits the parallelism to a single bus. The memory separation is still used to advantage in microcontrollers, as the program is usually stored in *non-volatile memory* (program is not lost when power is removed), and the temporary data storage is in *volatile memory*. Non-volatile memories, such as *read-only memory* (ROM) are used in both types of systems to store permanent programs. In a desktop PC, ROMs are used to store just the start-up or bootstrap programs and hardware specific programs. Volatile *random access memory* (RAM) can be read and written easily, but it loses its contents when power is removed. RAM is used to store both application programs and data in PCs that need to be able to run many different programs.

In a dedicated embedded computer, however, the programs are stored permanently in ROM where they will always be available. Microcontroller chips that are used in dedicated applications generally use ROM for program storage and RAM for data storage. Memory technology is crucial to the design and understanding of embedded computers, and Chapter Four is dedicated to this important topic.

Microprocessor/Microcontroller Basics

There are three groups of signals, or buses, that connect the CPU to the other major components. The buses are:

- Data bus
- Address bus
- Control bus

The *data bus width* is defined as the number of bits that can be transferred on the bus at one time. This defines the processor's "word size." Many chip vendors define the word size based on the width of an internal data bus. For the purposes of this book, however, a processor with eight data bus pins is an 8-bit CPU. Both instructions and data are transferred on the *data bus* one "word" at a time. This allows the re-use of the same connections for many different types of information. Due to packaging limitations, the number of connections or pins on a chip is limited. By sharing the pins in this way, the number of pins required is reduced at the expense of increased complexity in the external circuits. Many processors also take this a step further and share some or all of the data bus pins to carry address information as well. This is referred to as a *multiplexed address/data bus.* Processors that have multiplexed address/data buses require an external address latch to separate and hold the address information stable for the duration of a data transfer. The processor controls the direction of data transfer on the data bus.

The *address bus* is a set of wires that are used to point to the memory or I/O location that is to be read from or written to. The address signals must generally be held at a constant value for some period of time before, during, and after the data is transferred. In most cases, the processor actively drives the address bus with either instruction or data addresses.

The *control bus* is an assortment of signals that determine what kind of information is on the data bus and determines where the data will go, in conjunction with the address bus. Most of the design process is concerned with the logic and timing of the control signals. The timing analysis is primarily involved with the relative timing between these control signals and the appearance and disappearance of data and addresses on their respective buses.

Microcontroller CPU, Memory, and I/O

The interconnection between the CPU, memory, and I/O of the address and data buses is generally a one-to-one connection. The hard part is designing the appropriate circuitry to adapt the control signals present on each device to be compatible with that of the other devices. The most basic control signals are generated by the CPU to control the data transfers between the CPU and memory, and between the CPU and I/O devices. The four most common types of CPU controlled data transfers are:

1) CPU reads data/instructions from memory   *(memory read)*

2) CPU writes data to memory   *(memory write)*

3) CPU reads data from an input device   *(I/O read)*

4) CPU writes data to an output device   *(I/O write)*

In this book, "read" and "input" will be used interchangeably. These terms refer to the transfer of information from an external source into the CPU. "Write" and "output" will be used to denote the transfer of data from the CPU to an external destination. The data direction is defined with respect to the CPU.

Design Methodology

The address decode and control logic shown in Figure 2-2 is the key part of the design, which requires attention to timing analysis to guarantee signal logic and timing compatibility between the other blocks. The simplified timing diagram for such a system is shown in Figure 2-3. Figure 2-3 is a generic diagram and represents a typical example of a bus cycle for a typical CPU.

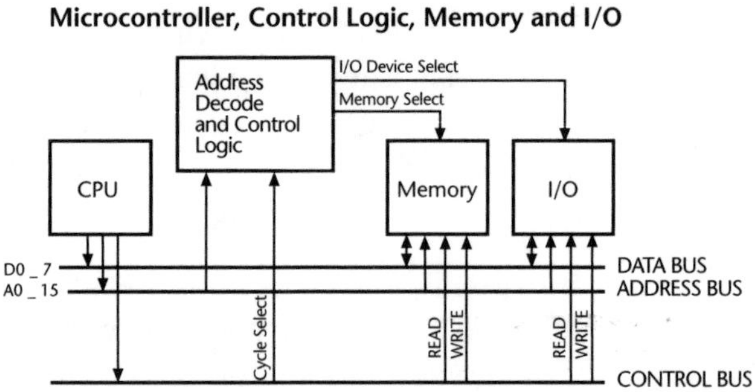

Figure 2-2 (right): Microcomputer busses.

Figure 2-3 (below): Generic bus timing example.

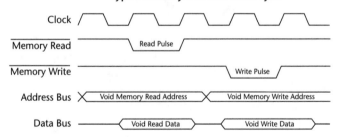

We see that there are two cycles:

- **Memory Read.** The processor places an address on the address bus, and activates the memory read signal by pulling it low, which causes the selected memory location to be placed on the data bus.

- **Memory Write.** The processor places an address on the address bus, data to be written on the data bus, and activates the memory read signal by pulling it low, which causes the selected memory location to be loaded with the data the CPU placed on the data bus.

Up to this point, we have discussed microcontroller architecture in a very general form, as it applies to most common devices. In order to go deeper into the operation of a microcontroller, it is appropriate to present one specific processor as an example. In order to really understand and apply this information to a real hardware and software design, it is necessary to cover one specific machine architecture in detail. That is what we will do in the next section.

The 8051 Family Microcontroller Processor Architecture

You might wonder why the 8051 family of processors was chosen for this purpose, as it is a relatively old processor. If you read current technical journal articles, you might get the impression that all the action is in 32-bit micros. That is primarily due to the fact that the companies that sell the high-end devices are working very hard to put their newest technology in front of their customers, and they are the ones who write most of the trade articles.

It is important to note that the trade press is always emphasizing the high end 16-bit, 32-bit, and larger processors due to their dependence on the advertising revenue from chip vendors. Though you would never guess it from reading these publications, it is only recently that shipments of 8-bit microcontrollers have exceeded 4-bit units. It will be quite some time before the 16-bit micros will approach the sales volume the 8-bit units have reached, and the 8-bit units are still growing in volume. According to one of the leading industry publications, there are more 8051 derivative CPU chips being produced than any other 8-bit micro. From this point forward, the 8051 family architecture will be used. Later on, other architectures and generic features not implemented in the 8051 will be discussed for completeness. Once you have learned the concepts of the 8051, you will find that the next architecture you need to use will be much easier to learn.

The 8051 microcontroller was chosen as the example processor in this book for several reasons:

- The timing specifications are simple and allow a complete detailed timing analysis within the limited scope of this book.
- Interfacing to the processor's multiplexed address/data bus provides valuable design experience.

- Development tools, including assemblers, simulators and compilers are readily available as freeware shareware and demo versions.
- It is available at a low cost, allowing low cost versions of in-circuit emulators, peripheral components, and single board computers to be purchased by the student.
- The 8051 is the most popular microcontroller family, with many derivatives available, and multiple vendors manufacture it.
- The 8051 architecture is available in a wide range of cost, size, and performance. For example, one version is available in a 20-pin small outline surface mount package for less than a dollar in volume, and another one is about eight to ten times the speed of the original 8051.
- The 8051 CPU is also available as a building block for custom chip designs, and is the most popular CPU for "system on a chip" designs. It is also the only readily available, non-proprietary building block CPU architecture available for chip design.

Software tools for the 8051 family, such as assemblers, compilers and simulators are available at no cost on the internet. Hardware tools, such as the combination software development kit and in-circuit emulator (the SDK which can be used in conjunction with this book), are available for under $100, and complete design documentation is available on the web to allow anyone to build their own.

In addition, the 8051 has the simplest timing specifications of a device which can address external memory, making it practical to go into the details of the design which are necessary to understand. With less than two dozen timing specifications (compared to several times as many for most other equivalent processors), it is possible to cover the timing specifications in detail. Once this process is understood, it is a straightforward jump to understanding and using the larger number of equivalent specifications characteristic of other devices.

Introduction to the 8051 Architecture

This section is intended to provide a broad overview of the 8051 microcontroller architecture. References to "8051" or "'51" in this book generally indicate the entire family of 8051 CPU instruction set compatible devices. Since the original 8051 had an internal read-only memory for programs—which was defined at the time the chips were fabricated—that device is not appropriate for our study. For flexibility and simplicity, we will be discussing the 8031, which does *not* have any internal program memory but instead fetches its program from an external memory device. Otherwise, almost all the versions of the processor family share the same features. If one were to do a practical commercial embedded computer design using an 8051 derivative, one could take advantage of the additional features that are commonly included in the more recent devices. For example, the NMOS versions of this family (e.g. 8031) described here have mostly been displaced by their CMOS counterparts, such as the 80C31. The 8032 and 80C32 with 256 bytes of internal data RAM and an additional timer, at about the same cost, have replaced the '31 versions. Most of the new versions of these devices have been built upon the features of the '32 version. Higher speed versions of the device, such as the Dallas Semiconductor 80C320, provide throughput equivalent to almost 100 MHz, compared to the original parts 12 MHz clock. The 8051 CPU element is even

available as a standard building block for use in designing other chips. There are also 16-bit superset versions of the 8051 architecture! A simple 8051 system is shown in Figure 2-4.

Figure 2-4 shows a highly simplified version of the CPU with external program and data memory. (An address latch is also required, but not shown in this figure.) The program is

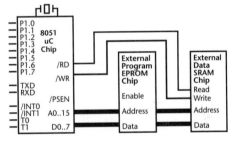

Figure 2-4: A simple 8051 system using external memories.

stored in non-volatile ROM memory, such as an EPROM (*erasable and programmable read-only memory*), and the data is stored in a volatile RAM. In this configuration with external memory, the amount of useable I/O is limited by the number of pins that are used for the address, data, and control lines. Only Port 1 and part of Port 3 is available for user I/O in this case. In its simplest configuration, only the processor's internal memory is needed for the application, so most of the pins are available for I/O. In that case, the microcontroller is the only required chip, which is also the lowest cost configuration. There are versions of this device that have internal program memory that can be programmed with an inexpensive programmer connected to a PC.

Now that we've introduced the 8051 architecture, we need to get into the "low level details" in order to really understand it. Up to this point we've had a view from 50,000 feet, where all the landscaping looks perfectly manicured.

Now we need to get down to ground level where we can see all the bits of trash and imperfections of reality. Every processor has its own idiosyncrasies, and the 8051 is no exception. While it may seem quite odd at first, it does have some very useful features which make it fairly adept at handling the sorts of things that are often required in an embedded application. Having said that, let's get down to looking at the innards of the processor. Figure 2-5 shows a top view of the processor with pin numbers, starting with pin 1 in the upper left corner.

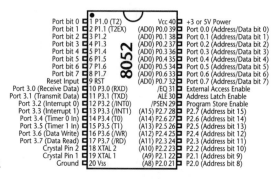

Figure 2-5: Top view of 8052 40-pin DIP package.

Figure 2-5 shows the pin numbers, names and functional description of the pin functions for the 8052 CPU in a *dual in-line plastic* (DIP) package. The 80x1 and 80x2 pin definitions are identical, except for the fact that the 80x1 does not have Timer 2, so those pins are different on the 80x1 parts.

8051 Memory Organization

In order to understand the processor, it is necessary to see how the various memory spaces are organized. The memory organization of the 8051 family of processors may seem complex at first; however, it as not as random as it might seem. There are separate memories for program storage, internal memory and registers, internal I/O functions, and external data memory. The program and external data memories are relatively simple. They each hold up to 64 kilobytes of instructions and data respectively. Program instructions are always fetched from program memory, and are indicated by the CPU activating the /PSEN pin. External data is transferred when the CPU executes a MOVX (MOV eXternal memory) instruction, and the CPU indicates this by activating the /RD or /WR line. The 8051 family chips only have three types of external memory cycles:

- Program read when /PSEN goes low
- External data read when /RD goes low
- External data write when/WR goes low

This makes interfacing other bus-oriented devices to the processor relatively easy. (Some general purpose or PC CPUs have many different types of bus cycles.)

The internal data address space of the 8051 family is not quite as simple as the external memories. It includes four banks of eight registers, memory that can be accessed one byte or one bit at a time, a stack, and the *special function registers* (SFRs) which hold the data and control information for the serial port, timers, and other I/O. This internal memory address space can be accessed in several different ways. The internal data space of the CPU can be rather confusing at first, but it is one of the characteristics of the 8051 family, which allows so much to be done with such limited resources.

The 8051 CPU manipulates operands in three memory address spaces:

- **64 kilobyte program memory** (external program memory on the 8031) which is enabled when the processor is fetching an instruction to be executed and signaled by activating the CPU's /PSEN control line. The MOVC instruction also activates /PSEN to enable reading the code memory into the accumulator for accessing lookup tables and other unchanging data stored in the program memory space.

- **64 kilobyte external data memory** which is enabled when the processor reads or writes data from the external data memory and signaled by activating the /RD and /WR control lines. This occurs only when a MOVX instruction is used to read or write from external memory.

- **Internal data RAM** (128 bytes for the '31, 256 bytes for the '32) and special function registers (SFR). Four register banks (each bank has eight registers), 128 individually addressable memory bits, and the stack all reside in the internal data RAM. The stack depth is limited only by the available internal data RAM. Its location is determined by the 8-bit stack pointer. The 128 byte special function register address spaces are shown in Figure 2-6.

Figure 2-6: 8031 memory address spaces.

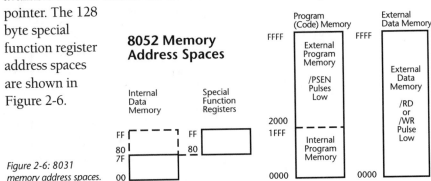

The lower 128 byte half of the 256 byte internal data memory address space contains four blocks of eight CPU registers, R0-7. In the 8032 CPU, the upper 128 bytes of the internal data memory address space are shared between data memory and the SFRs, depending upon the address mode. The upper 128 bytes of data memory must be accessed using the indirect register 0/1 (@R0 or @R1 operands) or stack accesses, and all other references to addresses of 128 or higher will access the SFRs. All registers except the program counter and the four 8-register banks reside in the special function register address space. These memory mapped registers include arithmetic registers, pointers, I/O ports, and registers for the interrupt system, timers and serial channel. There are 128 bit locations in the SFR address space that are addressable as bits. The 8031 contains 128 bytes of internal data RAM and 20 special function registers (SFRs), while most other processor family variants include an additional 128 bytes of internal data memory overlapped with the SFR addresses.

8051 CPU Hardware

The 8051 is classified as an 8-bit machine, since the internal ROM, RAM, special function registers, arithmetic/logic unit and external data bus are each eight bits wide. The 8031 is identical to the 8051, except that it does not have any internal program ROM. The 8051 performs operations on bit, nibble, byte and double-byte data types. It excels at bit handling since data transfer, logic and conditional branch operations can be performed directly on the bit addressable SFRs.

This section describes the hardware architecture of the 8051 CPU. A detailed 8051 functional block diagram is displayed in Figure 2-7.

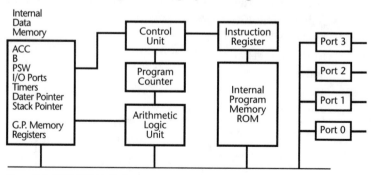

Figure 2-7: 8051 CPU block diagram.

Control Unit

Each program instruction is decoded by the control unit, which is also called the *instruction decoder*. This unit generates the internal signals that control the functions of all the other units within the CPU section. All instructions are fetched from the program memory ONLY. Instructions can be fetched from either the internal program memory (for those devices which possess one) or from external program memory. Instruction fetch operations are indicated when the CPU activates (lowers) the /PSEN line (NOT program strobe enable). A program memory fetch cycle lasts as long as /PSEN stays low. External program memory must only drive the data bus with the addressed instruction while /PSEN is low.

Program Counter

This is the pointer to the next instruction to be executed. The 16-bit program counter (PC) controls the sequence in which the instructions stored in program memory are executed.

Instruction Register

This is the register that contains the instruction that is currently being executed.

Internal Program Memory

The 8051 family has 16 address lines, and can directly address $2^{16} = 64$ kilobytes of program memory. The original 8051 has 4 kilobytes of program memory resident on-chip, the 8031 has no on-chip program memory, and the 8052 has 8 kilobytes of program memory. Other variants of the family are available with 1 to 64 kilobytes of various types of non-volatile program memory built in. The 64 kilobyte program memory address space is composed of a combination of internal and external program memory (external program memory only on the 8031 and 8032). When external program memory is accessed, and the processor is fetching an instruction to be executed, the external program read cycle is signaled by activating the CPU's /PSEN control line. The MOVC instruction also activates /PSEN to enable reading the code memory

into the accumulator for accessing lookup tables
and other unchanging data stored in the program
memory space. Figure 2-8 shows a program
memory map.

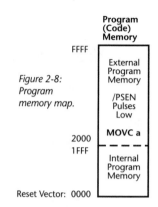

*Figure 2-8:
Program
memory map.*

The processor can fetch instructions from internal
or external program memory. There is a control
input pin, /EA (external access), which forces all
instructions to be fetched from the external
program memory when the pin is pulled low.
If the /EA pin is pulled high, then the processor
will fetch instructions from any available internal
program memory. When the processor first powers up and receives a reset
signal, it begins by executing the instruction at location 0000 in program
memory. When the processor fetches instructions from external program
memory, it puts the instruction address out on the address bus, pulses the
/PSEN (program strobe enable) pin low to enable the external program
memory to place the instruction on the data bus to the processor.

The generic part numbering scheme is as follows:

• 8xxx: NMOS logic

• 8xCxx: CMOS logic

• 803x: No internal program memory

• 805x: Factory programmed internal ROM program memory

• 87xx: Internal user programmable EPROM program memory

• 89xx: Internal flash EPROM program memory

• 8xx1: 4 kilobyte internal program memory, 128 byte internal RAM

• 8xx2: 8 kilobyte internal program memory, 256 byte internal RAM

For example, the 80C32 used as the standard processor in the SDK board is a
CMOS part with no internal program ROM, and 256 bytes of internal data RAM.

Internal Data Memory

Figure 2-9 shows the data memory spaces in the 8051. The internal data RAM
provides a convenient 128 byte scratch pad memory that includes the register

banks, SFRs, and general-purpose data storage. The programmer (or compiler) may also use this scratch pad memory for storing intermediate calculations on a temporary basis. The 8031 contains a 128 byte internal data RAM (addresses 0-7Fh, which includes registers R0-R7 in each of four banks), in addition to the memory-mapped special function register (locations 80-FFh). The 8032 has an additional 128 bytes of internal data RAM also at locations 80-FFh, which can only be accessed by using indirect register addressing (@R0, @R1) and the stack. The lower 128 byte half of the 256 byte internal data memory address space contains four blocks of eight CPU registers, R0-7. In the 80x2 CPU, the upper 128 bytes of the internal data memory address space are shared between data memory and the SFRs, depending upon the address mode. The upper 128 bytes of data memory must be accessed using the indirect register 0/1 (@R0 or @R1 operands) or stack accesses, and all other references to addresses of 128 or higher will access the SFRs. All registers, except the program counter and the four 8-register banks, reside in the special function register address space. These memory mapped registers include arithmetic registers, pointers, I/O ports, and registers for the interrupt system, timers and serial channel. There are 128 bit locations in the SFR address space that are addressable as bits. The 8031 contains 128 bytes of internal data RAM and 20 special function registers (SFRs), while most other processor family variants include an additional 128 bytes of internal data memory overlapped with the SFR addresses.

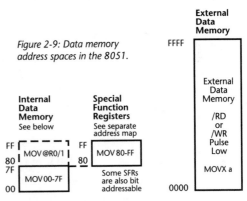

Figure 2-9: Data memory address spaces in the 8051.

Data Memory

The 8051 family devices have two data memories, internal and external. With 16 address bits, there is a maximum of 64 kilobytes of external data memory, which is useful for storing large blocks of variable information that will not fit in the internal data RAM. It is enabled when the processor reads or writes data from the external data memory, signaled by activating the /RD and /WR control lines. This occurs only when a MOVX instruction is used to read or write from external memory.

The internal data address space has two different parts, as shown in Figure 2-10. One part contains the general-purpose registers and general-purpose data storage RAM, and the other part contains all the special registers and I/O devices, such as the parallel and serial ports, and timers. These registers are called special function registers. There is a maximum of 256 bytes of internal RAM (128 bytes for the '31/'51, 256 bytes for the '32/'52) and special function registers (SFR). Four register banks (each bank has eight registers), 128 individually addressable memory bits, and the stack all reside in the internal data RAM. The stack depth is limited only by the available internal data RAM. The 8-bit stack pointer determines the stack's location.

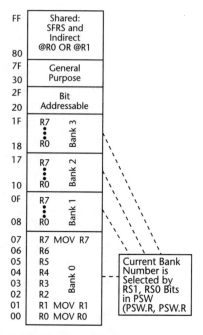

Figure 2-10: The internal data memory.

The internal data RAM provides a convenient 128 byte scratch pad memory which includes the register banks, SFRs, and general purpose data storage.

RAM locations 00-7F hex

- **Register banks:** There are four register banks within the internal data RAM. Each register bank contains registers R7-R0.
- **128 addressable RAM bits:** In the 8031, there are 128 addressable software flags in the internal data RAM. They are located in the 16 byte locations starting at byte address 20h and ending with byte location 2Fh of the RAM address space.

Special Function Register (SFR) locations 80-FF hex

- **General registers A, B,** and other registers are mapped here.
- **Parallel I/O ports:** The 8031 has four 8-bit ports.

- **Serial I/O port:** The serial I/O port built into the 8031.
- **Timer/counters:** There are counters that can count external events or count processor clock cycles to operate as timers. Many of the SFRs are also bit addressable.

Bit Addressable Memory

Figure 2-11 shows the organization of bit addressable space in the internal data memory. The bit address space has a total of 256 possible bit addresses. The first 128 bits, 00 to 7F hex, are used to access individual bits of the internal memory from location 20 to 2F hex. The second 128 bits, from 80 to FF hex, allow selected bits in the special function registers to be accessed at the bit level. Not all SFRs are bit addressable, and not all bit addresses are used in most processors.

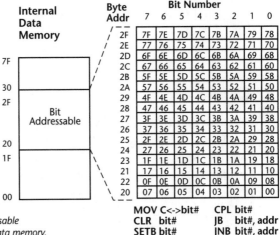

Figure 2-11: Bit addressable space in the internal data memory.

Bit addressable memory allows the manipulation and test of individual bits, which is a very common operation in embedded systems. Almost every application requires that some output bits be used to control an on/off device, such as an indicator or relay. Likewise input bits are used to sense the status of some external device, such as a switch or sensor. The bit addressable address space allows the programmer to operate on information at the bit level just as easily as at the byte level. This is contrasted by most other processors, in which the programmer must write multiple instructions to select the appropriate bit in a byte before processing or testing it.

Internal memory locations from 20 to 2F hex, are accessible either one byte at a time, or one bit at a time. That makes it easy to convert inherently serial information to parallel and vice versa, and to perform Boolean logic functions. This bit-level processing is one of the most unique and powerful features of the 8051 family architecture, and is one of the features that differentiate it from other microcontrollers.

Register Banks

The four register banks within the internal data RAM each contain eight registers named R0-R7.

128 Addressable Bits

There are 128 addressable software flags in the internal data RAM. They are located in the 16 byte locations starting at byte address 20h and ending with byte location 2Fh of the RAM address space.

I/O Ports

There are four 8-bit ports. When using external program or data memory, only Port 1 (P1) is available for general purpose I/O. External memory uses Port 0 (P0) for the multiplexed data bus and address bits 0-7, and Port 2 (P2) for address bits 8-15, while Port 3 (P3) contains special control signals, such as the read and write strobe pins. In addition to the basic parallel I/O bits on the four ports, some of the port bits have alternate functions. The alternate functions include the serial I/O port signals, timer and interrupt inputs.

Timer/Counter

The 8031 has two timer/counters and the 8032 has three.

Serial I/O

The serial I/O port that is built into the 8031 can be used to transmit and receive asynchronous (un-clocked) serial data, as is used on a PC's serial port. It can also be used for synchronous (clocked) serial data transfers.

Reset Circuitry

The reset input pin should be connected to an external resistor and capacitor, so that the processor will be properly initialized upon initial application of power. There is a capacitor between the reset pin and the power supply, and a resistor from the reset pin to ground.

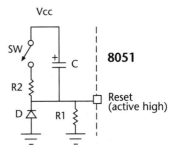

When power is first applied, the capacitor has no voltage across it, forcing the processor to reset. After resistor R1 charges the capacitor C, the reset signal goes low (inactive), and the processor begins executing the program beginning at location 0 in program memory. The recommended reset circuit is shown in Figure 2-12.

Figure 2-12: Recommended reset circuit for the 8051.

When power is first applied, capacitor C has zero voltage across it, and reset is held high until the current that flows through R1 charges C. Once the capacitor is charged, the reset pin is at zero volts and inactive. The diode allows the capacitor to discharge when Vcc goes to zero, even for a short period. If there was no diode, and there was a brief power loss, the CPU state would be indeterminate, and would not be reset. Optionally, the processor can be reset by closing switch SW through a series resistor R2, which limits the current through the switch. The current flowing through the switch discharges the capacitor. If resistor R2 was not present, very high currents could flow through the switch. These high currents that flow very briefly while the capacitor is shorted and can cause the switch contacts to fail or even weld the contacts together.

The R1*C time constant must be long enough to guarantee that the processor will be completely reset to a known state upon power up. The delay must allow the oscillator to start up and stabilize, as well as the time it takes the processor to reset after the oscillator is stable. Different processors require different numbers of clock cycles to reset themselves, and the oscillator start-up time can vary widely depending on the frequency reference, voltage, capacitive loads, and other factors. If the processor reset is not long enough, the processor may behave in unpredictable ways, and it may not be apparent that the problem is due to an incomplete reset operation. In most cases, it's better to have a relatively long reset time constant, on the order of hundreds of milliseconds, to be sure that the processor has been completely reset. External peripherals can also exhibit this problem. During the initial development of the SDK, we experienced occasional problems with the external serial port chip used on the board. The problem turned out to be related to the length of the reset pulse and the period of time *after* the reset when the chip must be left alone to pull itself together! This sort of problem can be very difficult to trace down, since it is difficult if not impossible to determine when a chip has not been completely reset.

The 8051 is unique in that its reset signal is active high. Other processors use active low reset signals, so the reset circuit must be adjusted to perform the equivalent function with the reset pulse going low at power up and when the capacitor is charged, the reset goes high. The circuit configuration except R and C1 are swapped, as are D1 and the SW/R2 pair.

The circuit in Figure 2-12 is good enough for most applications. However, it is not foolproof. Even with the above precautions, it is possible that the processor state can be jumbled by power transients that are too short to cause a reset. When a processor is used in a critical or long term unattended application, that probably won't be good enough to meet the need for reliable operation. To deal with this, processor supervisory chips are available to monitor the power supply voltage for out of tolerance fluctuations and automatically reset the processor when the power supply falls out of tolerance. Some of these supervisory chips also have a special "watchdog" timer circuit that expects to be "fed" by a pulse that resets the watchdog counter periodically by a correctly functioning program running on the processor. If the watchdog timer is not "fed" with a pulse periodically, the counter will overflow and it will "bark" by pulling the reset pin active. That way if the processor goes off in the weeds, due to a hardware glitch or a program bug, the CPU will be reset. This is a simple method of obtaining tolerance to fault conditions, but it also requires careful design to avoid undesired reset conditions. It is also the designer's responsibility to assure that the processor can't get stuck in a loop while feeding the watchdog timer.

When designing a microcontroller that must operate in high noise environments, or where correct operation is safety critical, special care must be taken to ensure that electromagnetic noise does not cause problems. This noise can come from other parts of the system and environmental conditions such as electromagnetic fields from other devices such as wireless communication devices. With the rapid increase in the number of electronic and wireless devices, this problem is becoming more and more serious. The field of *electromagnetic compatibility* (EMC) covers this noise, as well as others such as *electrostatic discharge* (ESD). A good summary of EMC concepts as they relate to micro-controllers can be found in the Intel application note AP-125, "Designing Microcontroller Systems for Electrically Noisy Environments."

Oscillator and Timing Circuitry

Timing generation is completely self-contained on the 8051, except for the frequency reference (which can be a crystal or external clock source). The on-board oscillator is a parallel anti-resonant circuit with a frequency range of 1.2 MHz to 12 MHz for the original 8051. There is a divide-by-12 internal clock counter that gives the standard 8051 an instruction cycle of 1 μS with a 12 MHz crystal. Higher speed versions of the processor are also available, which use fewer than twelve clocks per instruction cycle. The Dallas 80C320 uses only four clock cycles for most instruction cycles, so it is three times faster than the original CPU using the same clock frequency. The XTAL2 pin is the output of a high-gain amplifier while XTAL1 is its input. A crystal connected between XTAL1 and XTAL2 provides the feedback and phase shift required for oscillation. For stability and consistent oscillator start-up, two capacitors in the range of 10 to 20 picofarads should be connected from the XTAL pins to ground. If XTAL1 is being driven by an external frequency source, XTAL2 should not be connected. An external clock can also be applied to XTAL1 to allow the use of a separate clock frequency source, such as an oscillator module. Figure 2-13 shows a standard oscillator configuration.

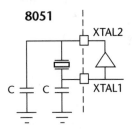

Figure 2-13: Standard oscillator configuration.

The oscillator circuit consists of a crystal connected between the XTAL1 and XTAL 2 pins of the processor, along with two capacitors, one from each XTAL pin to ground to improve stability and start-up characteristics of the oscillator. The internal amplifier and quartz crystal form a series resonant oscillator which operates at the at the crystal's resonance frequency. The amplifier in the original 8051 was an inverting amplifier, but other variants and other processor families make use of non-inverting amplifiers in some cases. All of the processor's timing is derived from this oscillator. For the standard 8051 compatible parts, each instruction cycle requires a multiple of 12 clock cycles. For the Dallas high-speed CPU versions, four clock cycles are used for most instruction cycles.

In most 8051 designs, the capacitors connected to the crystal should be in the 10 to 50 picofarads range, with 30 picofarads being a typical value. The crystal should be an "AT cut" series resonant device. The "AT" designation refers to the way the quartz crystal is cut from the blank with an orientation relative to the crystal lattice that reduces the crystal's frequency dependence on temperature changes. The crystal is manufactured so that it is series resonant at the specified frequency. A given crystal will resonate in a series or parallel mode. A parallel resonant crystal will still operate in the circuit, but it will operate at a slightly different frequency. Actual operating frequency depends on the load capacitance, and is subject to temperature, and will drift over time.

Selection of the capacitors is a trade-off between oscillator start-up time and stability. Specification of a crystal depends upon the specific design requirements and the processor being used. Even parts with the same number may have different requirements, especially for parts from different manufacturers.

There's much more information available from the crystal and processor manufacturers on the proper design and operation of crystal oscillators. Other frequency references, such as ceramic resonators and even simple R-C circuits can be used for many processors. Some microcontrollers even include on-chip oscillators that can be calibrated to operate at a specific frequency, albeit with less accuracy and greater drift. Application note AP-155, "Oscillators for Microcontrollers" from Intel Corporation, is a very useful reference and describes the characteristics of both the crystal and ceramic resonator's operation as well as the processor's oscillator amplifier.

The 8051 Microcontroller Instruction Set Summary

The following description of the instruction set is not a complete list, but serves to introduce the general character of the standard 8051 instructions. The instruction set utilized by the 8051 microcontroller consists of a total of 111 instructions, which may be divided up into several different categories. These are:

1. Arithmetic (24)
2. Logical (25)

3. Data transfer (28)

4. Bit (Boolean) variable manipulation (17)

5. Program branching and control (17)

Each of these categories is comprised of instructions that utilize *mnemonics* as shown below:

Arithmetic

ADD, ADDC, SUBB, INC, DEC, MUL, DIV, DA

Logical

ANL, ORL, XRL, CLR, CPL, RL, RLC, RR, RRC, SWAP

Data Transfer

MOV, MOVX, MOVC, PUSH, POP, XCH, XCHD

Bit (Boolean) Variable Manipulation

CLR, SETB, CPL, ANL, ORL, MOV, JC, JNC, JB, JNB, JBC

Program Branching and Control

ACALL, LCALL, RET, RETI, AJMP, LJMP, SJMP, JMP, JZ, JNZ, CJNE, DJNZ, NOP

Direct and Register Addressing

While the number of mnemonics is clearly smaller in number than the total of 111 instructions, a given mnemonic may be used in several different ways to make up a valid 8051 instruction. These different ways of forming instructions are classified by the types of *arguments* that a given mnemonic takes. A mnemonic can refer to data in a number of ways. One can refer to data located in particular address in the data memory space either by specifying its address directly, or indirectly by using a *data pointer register*. In this case, the data pointer register contains the address of the memory location we seek. The 8051 looks in the data pointer register, and then retrieves the information located in the location referred to (or *pointed to*) by the data pointer. Additionally, the 8051 has 32 bytes of internal memory divided up into four register banks of eight bytes each. These register banks may be referred to in an 8051 instruction by either their direct address (which ranges between 00h and 1Fh), or by their register name, which is denoted by R0 through R7. When these memory locations are addressed by their register name, it is important to remember which register bank is currently in use. These register banks, numbered 0 through 3, are selected through two bits located in a special register called the *program status word* (PSW). The PSW contains a number of very important bits, which are used to indicate the current status of the processor. Note that because the registers R0 through R7 are located in the data memory space, they may be addressed either by the register name or by their direct address location. Consider the instruction:

```
MOV A,R3
```

This instruction takes the contents of register R3 and moves it (actually, the data is copied) to a register denoted by the letter "A", called the *accumulator*. The accumulator is the "working" register of the 8051, and is the register that is used in most all arithmetic and logical operations performed by the processor.

Assuming we are using register bank 0, the following instruction is identical to the instruction just shown:

```
MOV A,03h
```

Since register R3 is at internal RAM location 03h, the above instruction takes the data stored in RAM location 03h and moves it to the accumulator.

What is the difference between these two forms of saying the same thing? The first instruction is called *register addressing*, while the second instruction is called *direct addressing*. The reason for the difference in nomenclature is obvious, and while it may seem a bit pointless to dwell on the difference between these two modes, there is a significant difference in the way the 8051 deals with each type of addressing.

Looking in the *80C51-Based 8-Bit Microcontrollers Data Book* (publication number IC-20) published by Philips, the instruction MOV A,R3 takes up only one byte of program memory space, while the instruction MOV A,03h requires two bytes of program memory space. The reason the register mode instruction requires less program memory to store is that a reference to a register requires three bits to represent its address, and a reference to an arbitrary location in internal data memory requires 8 bits. Once a particular register bank is selected by setting the proper bits in the PSW, any register in that bank may be completely determined by only 3 bits (3 bits are required to distinguish eight possible locations). If we use direct mode to perform the very same operation, we now require 7 bits to completely determine the exact location out of 128 possible locations—thus, direct addressing instructions generally occupy more program memory space than register addressing instructions.

There are two other memory locations in the 8051 that may be addressed through register mode. These are the accumulator, which we have already seen is denoted by the letter "A," and the *data pointer*, which is actually two registers. The letters DPTR denotes the data pointer, and is a 16-bit quantity used for addressing locations in data memory external to the microcontroller itself. Since the DPTR is a 16-bit quantity, a total of 64 kilobytes of data may be addressed. This is, of course, the maximum data that may be accessed at any one time by the 8051.

The following instructions are examples of data movement instructions that utilize direct addressing:

```
MOV 24h,A     ;move accumulator contents to internal RAM
                 location 24h

MOV 7Ch,0Fh ;move location 0Fh contents to internal RAM
                 location 7Ch

PUSH 22h      ;PUSH location 22h contents onto the stack

POP 4Eh       ;POP the top of the stack into location 4Eh
```

The following instructions are examples of data movement instructions, which utilize register addressing:

```
MOV R0,49h   ;move location 49h to register R0

MOV R2,A     ;move accumulator contents to register R2
```

Note that in all instructions, the order of the memory locations in the instruction is always *destination, source.* The destination address appears first, followed by the source address.

The instructions PUSH and POP perform operations on a portion of memory called the *stack.* While not a separate memory space, the stack is located in the internal data memory portion of the 8051/52, and is structured as a *LIFO* (last in, first out) data structure.

The instruction:

```
PUSH 49h
```

takes the data stored in internal RAM location 49h and puts it onto the top (that is, the first available location) of the stack. Exactly where the top of the stack is situated is determined by the value contained in the stack pointer (SP) special function register. When the processor executes a PUSH instruction like the one above, it first increments the SP register by 1, and then copies the internal RAM register specified in the PUSH instruction to the address pointed to by the SP register. In other words, the value contained by the SP register is a pointer to the memory location one byte below the top of the stack.

The POP instruction takes the data at the top of the stack and copies it to the internal RAM location specified in the POP instruction. After copying the data, the SP is decremented by 1. The SP register in the 8051/52 is therefore a pre-increment, post-decrement register. In the 8051, which contains 128 bytes of internal data RAM, the maximum legal value that the SP register may contain is 07Fh. The 8052 has an additional 128 bytes of internal RAM, separate from the special function registers. This section of RAM is accessible through the stack, and so the 8052 permits a maximum value of the SP register of 0FFh.

The SP register can be set by the programmer to any value that is convenient for the particular application. When the processor comes out of RESET, the SP register is loaded with 07h, thus placing the top of the stack at internal RAM location 08h. This is just above register bank 0. The stack always grows *upwards* through internal RAM. Care must be taken that the stack does not collide with other registers in internal RAM that have other uses. Additionally, if the SP register reaches its maximum value, 0FFh, and then overflows, the stack will continue to grow through the Bank 0 registers. As no stack overflow or underflow features are present on the 8051, this becomes the responsibility of the programmer.

Indirect Addressing

In many applications, it is inconvenient or impossible to always refer to data directly or as a register. When large amounts of data are being manipulated, either in internal or external data memory, very often it is required to address such data through the use of a *data pointer*. Use of a data pointer to address data memory is known as *indirect addressing*. The 8051 has four different methods by which data may be addressed indirectly:

1. The indirect registers R0 and R1, located in each of the 4 register banks
2. The data pointer (DPTR) and the accumulator
3. The program counter and the accumulator
4. The XCHD instruction

Indirect addressing of data is used frequently. Many embedded applications require calculation of one form or another, and frequently the most efficient means of doing this is through the use of a *look-up table*. As an example, an 8051 microcontroller such as the 80C552 has an eight channel, 10-bit analog to digital converter (ADC). The ADC takes an analog voltage as its input, and converts it to a 10-bit binary number between 000h and 3FFh. If this ADC is used, for example, to convert the analog output voltage of a pressure transducer to a digital value, it is necessary to relate each of the 1024 possible counts of the ADC to a pressure value. If the computer in use is very fast, or has a great deal of floating point mathematical ability, it would be possible to directly calculate the pressure value from the ADC count—one would need the characteristics of the transducer to accomplish this. However, an 8-bit embedded controller such as the 8051 does not have such capability, or at least the ability to do complex mathematical calculations quickly. In this case, it is far more efficient to simply generate the 1024 numbers that correspond to the pressure output of the transducer and store these numbers in a table. The processor then takes the output of the ADC and uses this 10-bit number as an *offset* into the table stored in RAM. This offset, when added to the *base address* of the lookup table (the base address is the address of the first record in the table), constitutes the physical address of the data record that corresponds to the actual pressure sensed by the transducer. Since this lookup table could be located literally anywhere in either the code or data memory spaces, and since each record could be more than a single byte, it is in general not possible to store the actual location of each entry in the table. Rather, the ADC output is used to *indirectly address* the data through the use of a data pointer.

Registers R0 and R1 in each of the four register banks may be used to indirectly access any of the internal data memory space of the 8051. To illustrate by example, consider the instruction:

```
MOV A,@R1
```

Here, the "@" symbol is used to denote indirection, similar to the asterisk "*" in C. This instruction takes the data located in the location pointed to by register R1 and copies it to the accumulator. Note that the value copied to the accumulator is not the contents of R1, but the value in the memory location equal to the contents of R1. This is why register R0 is said to be a data pointer, pointing to another internal RAM location. Notice that only data located in the internal data memory space of the 8051 may be accessed through @r0 or @R1 instructions. As these registers are only eight bits wide, a total of 256 bytes may be specified. The 8051 microcontroller contains a total of 128 bytes of internal RAM located between addresses 00h and 7Fh, while the 8052 contains an additional 128 bytes of internal RAM between 80h and 0FFh. These upper 128 bytes of internal RAM can only be accessed by indirect addressing. It is important to distinguish these upper 128 bytes of internal RAM in the 8052 microcontroller from the special function registers. The SFRs are not part of the upper 128 bytes of internal RAM—they are a separate memory space.

Very often, an embedded system will require a much larger amount of RAM than is present on an 8051 or 8052 microcontroller. When this is the case, one generally uses external RAM chips that are interfaced to the processor over the address, data, and control bus structure. Since the address bus of the 8051/52 microcontroller family is 16 bits wide, a total of 64 kilobytes of either program memory or data memory may be accessed. Restricting our attention to the data memory space and RAM for the moment, we need some way of accessing the (at most) 64 kilobytes of RAM external to the microcontroller. The MOVX instruction (X denotes "external") is used to move data into and out of RAM located external to the microcontroller. The only way the 8051/52 microcontroller can access external RAM is through indirect addressing.

The MOVX instruction can be used in two different ways. If the external RAM space is small (small meaning 256 bytes or less in this case), it may be accessed with an 8-bit address. The R0 and R1 registers may be used in this manner just as they are used for indirect addressing of internal RAM. Consider the instruction:

```
MOVX @R0,A
```

This instruction takes the byte in the accumulator and copies it to the location at the address in external RAM pointed to by R0.

The instruction

```
MOVX @R1,A
```

performs the opposite operation. It takes the value held in the external RAM location pointed to by R1 and copies it to the accumulator. There is an important difference between this type of external data addressing and internal data addressing—whenever data is being read from or written to external RAM, either the source or the destination register must be the accumulator.

What if our external RAM array contains more than 256 bytes? Recall that the 8051/52 family of microcontrollers have a 16-bit address bus, permitting up to 64 kilobytes of external program and/or data memory. The *data pointer* (DPTR) is used to store a 16-bit address for indirect addressing of external RAM. DPTR is loaded with the address of interest, and the instruction

```
MOVX A,@DPTR
```

copies the data at the external RAM location pointed to by the 16-bit address pointer, called DPTR into the accumulator. The instruction

```
MOVX @DPTR,A
```

performs the opposite operation. The contents of the accumulator A is copied to external RAM at the location pointed to by DPTR. Again it is important to notice that either the source or the destination register in the instruction must be the accumulator.

Sometimes it is necessary to store information other than actual program instructions in a nonvolatile memory. Critical configuration data, lookup tables, or serial number information for unit identification oftentimes must be stored and available at system power-up without having to be regenerated by the system itself. While there are external nonvolatile memory technologies available (EEPROM and flash, for example), it is possible to use the program memory space of the 8051/52 for this same purpose. While it is not possible to *write* to the program memory space during normal operation (that could have potentially disastrous results!), it is possible to read data from it. The MOVC instruction ("C" denotes "Code") copies a byte in the program memory space to the accumulator. In order to accomplish this, the instruction requires the use of a base address and an offset. It is best to illustrate this with some examples. The two allowable forms of the MOVC instruction are:

```
MOVC A,@A+DPTR
```

```
MOVC A,@A+PC
```

In each of these instructions, the contents of the accumulator and either the DPTR or the PC (the program counter register) are added together, generating a 16-bit address. The contents of the address in the program memory space pointed to by this 16-bit sum is copied to the accumulator. In this way, either the PC or the DPTR can be used as a base address into a data table in the program memory space. The accumulator then becomes an offset into the data table, with a maximum offset value of 256.

The last method of indirect addressing available in the 8051 is the XCHD (exchange digit) instruction. The XCHD instruction is frequently used when BCD (binary coded decimal) arithmetic is being performed, or when a BCD lookup table is stored in internal RAM (a common use of a BCD lookup table would be for driving a 7-segment LED display). The XCHD instruction has the following syntax:

```
XCHD A,@R0

XCHD A,@R1
```

This instruction exchanges the low nibble (that is, the low 4 bits) of the accumulator with the low nibble of the internal RAM location pointed to by either the R0 or R1 register. Recalling that BCD uses 4 bits to represent the decimal numbers 0 through 9, this instruction offers a quick way to indirectly address a BCD (or any other 4-bit coding scheme) lookup table in internal RAM. To illustrate this with an example: suppose the accumulator contains A6h, register R1 contains 43h, and internal RAM location 43h contains 0BBh.

The instruction:

```
XCHD A,@R1
```

Will result in the accumulator containing 0ABh, and internal RAM location 43h containing 0B6h.

Immediate Addressing

Sometimes it is necessary to place a fixed constant into a memory location. This may be performed through the use of the immediate operator "#". As an example,

```
MOV A,#09h
```

places the number 09h into the accumulator. Likewise,

```
MOV 52h,#3Ah
```

places the constant 3Ah into internal RAM location 52h. The immediate operator indicates that the number that follows is to be interpreted as an *immediate constant*, rather than a memory location. Notice that, had we issued the instruction

```
MOV 52h,3Ah
```

this would have copied the contents of internal RAM location 3Ah to internal RAM location 52h. Since this is a perfectly valid 8051 instruction, the assembler will not flag this as an error if we had actually meant to prefix the 3Ah with the immediate operator. The code will not function as we might expect it to operate. **Watch out for this – it is a VERY common error!**

Immediate data, by its very nature, must only occur as the source operand of an 8051 instruction.

The instruction

```
MOV #52h,44h
```

makes no sense, and will be flagged as an error by the assembler. On the other hand, below is a valid instruction that will put the number 44h into internal data RAM location 52h:

```
MOV 52h,#44h
```

A detailed list of all of the instructions and their operations is contained in the 8051 programmer's reference handbook.

Generic Address Modes and Instruction Formats

Regardless of the type of processor, certain address modes are usually available in one form or another. This section describes some of the generic address mechanisms and instruction encoding formats, using the 8051 instructions and address modes as an example.

Instructions can be classified by the number of operand addresses that are explicitly specified. For example, "CPL A — complement accumulator" is an instruction that does not contain an explicit address, so it is a *zero-address instruction*. The accumulator is called an *implied operand* because the instruction op code does not have an address field, since this instruction always refers to the accumulator. An *explicit operand* has an address field embedded in the instruction op code or follows the op code, usually as a pointer to the data that is to be used. Other examples of zero-address instructions include PUSH, POP and RETurn because the operand is implied to be on the stack. The instruction "MOV A,address" — load accumulator with the content of internal memory location address" is a *one-address instruction* because the accumulator is an implied address, but the memory location is specified explicitly by its address. A *two-address instruction*, such as "MOV addr,@R0" (move the data at address pointed to by R0 to "addr") has two address fields. Some processors have three-address instructions, which allow the processor to perform an operation on two operands and store the result in a third operand, all of which may be referred to explicitly.

Instructions for a typical 8-bit CPU might consist of one or more op code bytes followed by optional operand fields. The first (op code) byte would identify the type of instruction, and the optional byte(s) following it would be the operand(s) or addresses of the operand(s).

8051 Address Modes

Implied addressing, as described above, always references the same location and does not have an explicit address field in the instruction. The instruction shown would take only one byte, and would result in only one memory cycle to fetch the op code byte. Table 2-1 illustrates this.

Instruction	Operand
CPL complement	A accumulator
E4 (op code)	A (implied)

Table 2-1:
Implied addressing.

Immediate addressing is used when the operand is a constant value, and is part of the instruction, usually immediately following the op code in program memory. An example would be an instruction that loads a constant into the accumulator, as shown in Table 2-2.

Instruction	Operand
MOV A, load accumulator	#35H with 35 hex
74 (op code)	35 (constant)

Table 2-2:
Immediate addressing.

The instruction would be stored in an 8-bit processor's memory as follows:

Address Value(hex)

1000   74 op code

1001   35 operand

Execution of this instruction would result in two memory cycles, one to fetch the op code and one to fetch the constant.

Direct addressing includes the address of the operand as part of the instruction rather than the operand itself. The address part of the instruction acts as a pointer to the data to be accessed. An instruction that loads the byte of data stored in memory location 1234 into the accumulator would consist of the op code followed by the address 1234.

Instruction	Operand
MOV A, load accumulator	34H with the contents of location 34
74 (op code)	35 (constant)

Table 2-3:
Direct addressing.

The instruction could be stored in an 8-bit processor's memory as follows:

Address Value(hex)

1000   E5 op code

1001   34 operand address

Execution of this instruction would result in three memory cycles, one to fetch the op code and one to fetch the address and one to fetch the byte at location 1234.

When dealing with values of more than eight bits, different microprocessor vendors use different methods of storing data in memory. Of course, Intel and Motorola chose opposite methods. The 16-bit address stored high byte first followed by the low byte as it is done in the Motorola 68000 family. Other processors, such as the Intel CPUs, reverse the order. For machines that support two byte or four byte values, there are two different ways of storing the bit operands: low byte first (Intel), and high byte first (Motorola).

Indirect addressing specifies a memory address that contains the address of the data to be transferred. An instruction that loads the byte of data that is pointed to by the address stored in memory location whose address (1234h) resides in the 16-bit register DPTR into the accumulator is shown below.

The instruction could be stored in an 8-bit processor's memory as follows, assuming that DPTR contains 1234h:

Instruction	Operand
MOVX A, load accumulator	@DPTR contains the address of the byte to be accessed
E0 (op code)	DPTR=1234h (address of the operand)

Table 2-3: Indirect addressing.

Code Address	Value(hex)	External Memory Data address	Value
1000	E0 op code	1234	57 operand

After completion of this instruction, the value 57 would be left in the accumulator. Execution of this instruction would result in two memory cycles, one to fetch the op code (E0), one to fetch the contents (57) of the address (1234).

The 8051 does not support true indirect addressing. In processors that do, the address of the operand is stored at the location contained in the instruction op code.

Register indirect addressing (e.g. MOV A,@R1) uses the contents of a register as an address, so the instruction would consist of only an op code byte. A register points to the operand in memory, so there is no need for an address field in the instruction. Two memory cycles are needed, one for instruction fetch and one for fetching the data.

Indexed addressing (e.g. MOVC A,@A+DPTR) is a combination of direct and register indirect addressing, because the instruction includes an offset address (DPTR), which is added to an index register (A register) to determine the address of the data to transfer.

It should be noted that the nomenclature for the various address modes varies, and the 8051 family address modes used for the examples above are not necessarily the best examples, as other processors support more extensive and flexible address modes.

The Software Development Cycle

The standard software development process consists of the following steps:

1) Create or edit an ASCII text file containing the human readable source code, also known as the program instructions.

2) Translate the source code to machine-readable binary instruction code using a language translator. This is accomplished using an assembler or compiler.

3) Load the program memory with the binary instruction code into the processor's program memory chip. For the SDK, the program is downloaded into program memory on the SDK.

4) Execute the program to test it and find program errors. For the SDK, this "debugging" process is facilitated using a special program (debugger or monitor) resident on the SDK.

5) Once the problem is located, the source code is corrected by repeating this process until all errors are corrected.

Software Development Tools

Software tools include translators, like assemblers and compilers, and debugging tools. Active debugging tools include:

• In-circuit emulators (ICE) for HW/SW integration; these are plugged into the application circuit (the "target" system) in place of the CPU, allowing the designer to "see inside" the microcontroller, download, and execute programs selectively.

- ROM emulators (ROM ICE) that allow the designer to reduce the time it takes to edit-compile-load-debug programs by replacing the program EPROM with a RAM that can be loaded quickly and easily from the host computer.
- Simple tools, such as an LED or speaker can also be useful in debugging.

Hardware Development Tools

There are two general classes of hardware development tools available to the embedded developer: passive analysis tools that allow looking at the operation of the system, and active tools. Active tools allow the designer to intrude on the operation of the system while it's running, even making changes to the system's configuration and software while it is under test. The system under test is usually referred to as the "target" system, and the computer that is used to develop, edit, compile, assemble, and download the code to the target system is called the "host" system.

Hardware tools include logic probes to display static logic levels and detect pulses, oscilloscopes to look at signal waveform amplitude vs. time, logic analyzers (with processor specific probes), and PROM programmers.

Chapter Two Problems

1. Processors such as the 8031 use multiplexed address/data buses. They require more than one clock cycle to do a memory transfer because some or all of the bus lines are shared. 16-bit addresses alternate with 8-bit data. The ALE (address latch enable) signal indicates when address information (A0-7) is present on the multiplexed address/data bus. The ALE signal is used to latch the least significant eight bits of the address in an 8-bit register. A minimum of two clock cycles is required to transfer data: one for latching the address when ALE is high, and one for the actual data transfer. How many clock cycles (minimum) would be required if the processor was a 16-bit machine doing a 16-bit transfer? Would the address latch have to be different?

2. How many unique locations could be referenced as "address zero" in the 8031 CPU architecture? (Remember to consider *all* the address spaces!)

3. Most processor control lines are active low. Comment on the reasons for this.

Worst-Case Timing, Loading, Analysis, and Design

Just as in comedy, timing is essential to the success of a microcomputer design. Often it is quite possible to get one system functioning by just interconnecting the various components. But it is *significantly* more difficult to be able to guarantee that many systems will work under the entire range of possible conditions that they may be exposed to. There are many designs in production right now that have a number of unidentified failures due to the lack of a worst-case analysis of the design. When timing or loading problems show up in a design, they usually appear as intermittent failures or as sensitivity to power supply fluctuations, temperature changes, and so on.

A *worst-case design* takes into account all available information regarding the components to be used with respect to variations in performance. Even when all parameters are at their most adverse values, the worst-case design can still be proved to meet the specifications. These variants may be due to changing manufacturing conditions, temperature, voltage, and other variables. Without performing a detailed analysis, there is no way of knowing if the design will work reliably under all operating conditions. It is much better to design reliability and simplicity of manufacturing into a product using worst-case design rules than to attempt to correct a problem after the design has been implemented. With the emphasis that must be given to the quality of the final product, a designer is obligated to perform a detailed examination of the timing in a system. As is the case in most quality improvements, these efforts result in direct cost and saving time. *This is clearly one of the places where the designer can have the greatest impact on overall product quality.*

Timing Diagram Notation Conventions

Timing notation is illustrated in Figure 3-1. The timing notation used in manufacturers' data sheets may vary from this, but is usually very similar. It is also important to notice that while the diagrams are reasonably standard, there is a wide variation in the selection of symbols for each timing parameter.

The purpose of timing analysis is to determine the sequence of events in each of the bus cycles so that we can delimit, among other things, the time available for each of the components to respond to changes. This time is compared to the requirements as specified in the manufacturers' data sheets to determine if they are compatible, and by what margin.

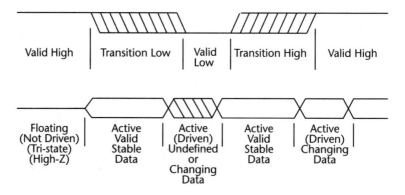

Figure 3-1: Timing diagram notation as used in this book.

The most important timing specifications for interfacing components to a bus-oriented design are:
• Rise/fall time
• Propagation delay time
• Setup time
• Hold time
• Tri-state enable and disable delays
• Pulse width
• Clock frequency

There are two general classes of logic: *combinatorial* and *sequential*. Combinatorial logic has no memory and its output is some logical function of its current inputs, after some delay. Examples of combinatorial logic include gates, buffers, inverters, multiplexers, and decoders. Sequential logic has memory, which means that its outputs are a function of both current and past inputs. Examples of sequential logic are flip-flops, registers, microprocessors, and counters. There are two types of sequential logic. *Synchronous* logic is synchronized to change only when there is a clock transition. In contrast, *asynchronous* logic does not use a clock signal. Almost all of the logic used in a microcomputer design will either be un-clocked asynchronous logic (gates, decoders) or clocked synchronous logic (counter, latch or microprocessor). Some types of devices are available in either form. Each of the timing specifications in the following discussion is described using simple logic devices as they are typically used in embedded computer designs.

Rise and Fall Times

The *rise time* of a signal is usually defined as the time required for a logic signal voltage to change from 20% to 80% of its final value. The *fall time* is from 80% to 20%, as shown in the figure below. These times are also commonly defined by some manufacturers as the transitions between the 10% and 90% levels. Figure 3-2 illustrates rise and fall times.

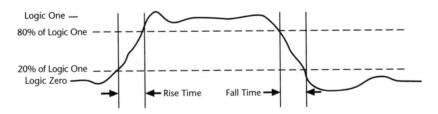

Figure 3-2: Rise and fall time of a signal.

Propagation Delays

The *propagation delay* is the time it takes for a change at the input of a device to cause a change at the output. All devices—even wires—exhibit some propagation delay. Some devices do not have symmetrical delays for positive and negative transitions. In the Figure 3-3, the propagation times for a high to low transition are shorter than for a low to high transition. This *asymmetrical delay* is common for TTL and open collector and open drain outputs because they are better at sinking current than sourcing it. Thus, the load capacitance is charged more slowly when the current is being supplied from the weaker "high side" or pull-up device. Propagation delays are usually measured from the 50% amplitude points, as shown in Figure 3-3.

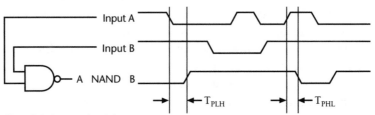

Figure 3-3: Propagation delay.

Setup and Hold Time

In Figure 3-4, a standard D type flip-flop (e.g., a 74xx74 device) is shown along with a sample timing diagram that illustrates the operation and key timing parameters of a flip-flop. This type of flip-flop samples the D input whenever the clock (CK) line goes high, and after a delay, the output remains in the same state until the next rising edge on the clock line. The triangle on the clock input indicates that it is a rising edge sensitive input, meaning that it will only have an effect when there is a rising edge on the clock pin. A falling edge sensitive input would have a bubble outside the block where the clock enters the flip-flop. In order to be able to guarantee that the flip-flop will operate correctly, the D input must be stable during the setup and hold time.

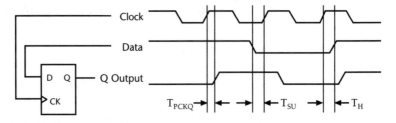

Figure 3-4: Setup and hold time.

Figure 3-4 also shows the propagation delay from clock to Q out (T_{PCKQ}), the setup time (T_{SU}), and the hold time (T_{H}). *Setup time* is the amount of time a sampled input signal must be valid and stable prior to a clock signal transition. *Hold time* is the amount of time that a sampled signal must be held valid and stable after a clock signal transition occurs. If these conditions are not met, the Q output may become invalid or even oscillate. This condition is referred to as *metastabilit*. The times of these and most other signals are frequently measured with respect to the 50% amplitude points of the clock signal rather than the valid logic one and zero levels. An analogy for the flip-flop as a sampling device is that of an instant camera: the clock is the shutter, the D input is the lens, and the output is the film image. The input is sampled when the shutter is open, and if the subject moves with the shutter open the picture will be blurred. For the flip-flop, the "shutter open" time, referred to as the *window of uncertainty*, is shown in Figure 3-5 below along with some possible results.

Metastability of a storage device such as a flip-flop or register is caused by the change of an input signal too close to the edge of the clock signal. In other words, if the setup or hold time requirements are not met, the output of the device is unpredictable and may even be unstable! The output may operate normally, take an invalid level, or oscillate (which may also explain why indecisive people take bad photos!)

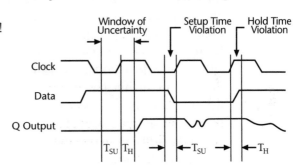

Figure 3-5: Metastability of a flip-flop.

Tri-State Bus Interfacing

When multiple devices are capable of driving the same line, the possibility exists that two or more of them will try to drive it in opposite directions at the same time. When tri-state devices fight like this it is called *bus contention*. Figure 3-6 illustrates this condition. While the data is unpredictable during this period, there are far worse things that can happen as a result of this condition. Since most tri-state devices have the ability to drive many loads, they are also capable of sourcing and sinking large currents. When two of these devices are in contention, very large currents with peaks in the tens or hundreds of amperes can flow for times on the order of nanoseconds.

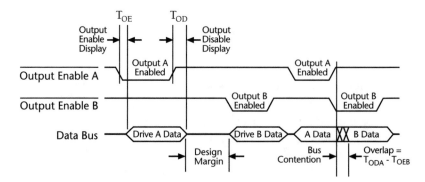

Figure 3-6: Tri-state bus timing and contention.

The large current spikes that occur during contention may stress the devices and significantly reduce their reliability. A far more frequent problem, however, is the temporary drop or glitch in the local power supply wires that can cause any other nearby devices to change state. As you can imagine, this can create havoc in sequential logic, particularly for micros. Based on past experience with Murphy's Law, these glitches generally seem to change the current instruction to "jump immediate to format hard disk routine," thereby erasing all your data. In a properly designed system, there is a "dead time" when no device is driving the bus to act as a safety margin between the times that two devices are enabled to drive their outputs. The problems arise when the output enable time of a device which is just turning on is less than the output disable time of a device which is turning off.

Pulse Width and Clock Frequency

The *width* of a positive going pulse is the period beginning from its *positive transition* (rising edge or leading edge) to its *negative transition* (falling or trailing edge). Figure 3-7 illustrates these concepts. Pulse widths are important in defining the operation of control signals such as the memory read or write signals and clocks. Clock signals used for modern microprocessors usually, but do not always, have equal high and low pulse width requirements. The period (T) of a signal is the sum of the rise time, high time, fall time, and low time. The frequency of a processor clock (f = 1/T) may have a lower limit as well as an upper limit. The standard NMOS 8051 family of parts has a lower frequency limit of 1.2 MHz. That means that the processor cannot be operated at a lower frequency. The reason is that the processor's internal design requires a constant clock, in order to correctly maintain its state. Other processors (such as the 80C51 series CMOS devices) can tolerate having their clock stopped completely, as they have been designed to maintain their internal states indefinitely, as long as power is applied.

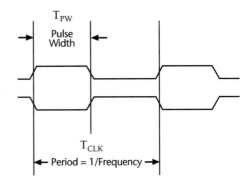

Figure 3-7: Pulse width, period, and clock frequency.

Fan-Out and Loading Analysis—DC and AC

Another important part of worst-case design is a realistic model of the signal loading for each of the circuit's outputs. If insufficient drive is available, buffer circuits must be added or the number of loads must be reduced to guarantee correct operation. *Fan-out* is the number of equivalent inputs that can be safely driven by one output. A fan-out of 10 indicates that one device output can drive ten inputs. The fan-out is determined from:

- The source, type and number of loads
- DC characteristics sources and load
- AC characteristics of the loads vs. the source test conditions

DC characteristics of the output and inputs consist of:

- The maximum current that can be produced by an output
- Maximum currents required to drive an input

The maximum output currents are specified as:

- I_{OLmin} Minimum output low (sink) current for a valid zero output voltage
- I_{OHmin} Minimum output high (source) current for a valid one output voltage

Note that a low output is sinking currents that are coming out of the inputs that are being driven. Likewise, a high output is sourcing current that goes into the inputs that are being driven.

Maximum currents required to drive an input are specified as:

- I_{ILmax} Maximum input low current for a valid zero input voltage
- I_{IHmax} Maximum input high current for a valid one input voltage

Another important convention has to do with the sign of the current flowing in or out of a device pin. In most cases, current flowing into a device pin is given a *positive* sign (as shown in Figure 3-8), while current flowing out of a pin is given a *negative* sign (as shown in Figure 3-9). In both Figures 3-8 and 3-9, the device on the left is the driving device, which tries to force its output to the desired logic state. In the logic one state, the output sources current (–50 microampere), and the receiving device absorbs that current (+50 microampere). In the example below, the available output current is exactly equal to the input current used by the load, resulting in a DC fan-out of 1.

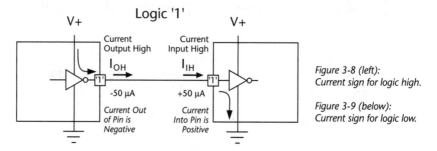

Figure 3-8 (left):
Current sign for logic high.

Figure 3-9 (below):
Current sign for logic low.

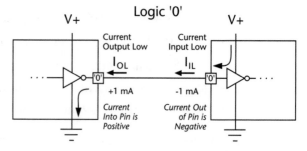

Unfortunately, **this convention is not always followed consistently,** so it is up to you to recognize the current direction from the context of the situation in which it appears. Generally, the current direction can be determined by keeping these images in mind, especially since many data sheets do not specify the sign for the input and output currents.

The other type of fan-out limitation is the ability of an output to drive the capacitance of the loads and stray wiring capacitance, also known as *AC fan-out*. The AC fan-out is determined by the specified test load for the driving chip, and the load presented by the actual load capacitance. The capacitive load is the parallel combination of all the input capacitances of the gate inputs attached to the signal, plus the wiring capacitance. Since the capacitors in parallel are equivalent to a single capacitor equal to the sum of the individual capacitances, we just add up all the load capacitor values and compare this to the output's specified test load. The driving device's specified load capacitance, C_L, the test load capacitance used by the manufacturer for specifying the AC or timing characteristics of the device. Most often, this specification is listed in the test conditions or notes for the timing specifications of the chip. As long as the sum of the load capacitances, including the stray wiring capacitance, is less than the specified test load for the driving device, all the timing specifications will be valid as specified in the timing section of the data sheet. If the driving device is overloaded (actual C_L is greater than specified C_L), then the timing specifications of the device need to be de-rated (slowed down), since additional capacitance will increase the rise and fall times of the signal line in question. Methods for estimating the amount that an overloaded output can withstand are described later.

AC characteristics of the outputs and the inputs consist of:

- C_L The load capacitance that an output is specified to drive, is listed in the timing specifications for the driving device under the name "test conditions" which is usually in the notes at the bottom of the specification sheet.
- C_{in} Maximum input capacitance of a driven input load.
- C_{stray} Wiring and stray capacitance can be approximated to be in the range of 1 to 2 picofarads per inch of wiring on a typical PC board.

As long as the inequality below is satisfied, the signal will meet the timing specifications for the driving device. If the actual load is greater, it will delay:

Driving device spec C_L > actual Cload = $C_{in1} + C_{in2} + \ldots + C_{wiring}$

The AC fan-out is limited by the parallel combination of the logic inputs' capacitance, C_{in}, and the stray or wiring capacitance. Capacitors in parallel are additive, so the load presented to an output is the sum of the input capacitances of the logic inputs plus the wiring capacitance. Logic input capacitance is often difficult to find, as it may not be listed in the component data sheet, but rather in another section of the data book describing the characteristics common to all members of a given logic family. Typical logic input capacitance ranges from 1 to 5 pF (picofarads or 10^{-12} F), but may be outside this range. The maximum load capacitance which a device is specified to drive (C_L), is usually defined in the test conditions for the timing specifications of an integrated circuit, as it is the timing which is most affected by capacitance. Load capacitance is usually specified in the range of 50 to 150 pF. Wiring capacitance is often in the range of 1 to 2 pF per inch of wire for a nominal printed circuit trace. Actual values can vary quite a bit, depending upon the physical dimensions of the trace, proximity to surrounding signals and distance from a ground plane, as well as the dielectric constant of the circuit board material.

Calculating Wiring Capacitance

The standard formula for determining capacitance is:

$$C = (\varepsilon * A)/d$$

Where A is the area of two closely spaced parallel plates, d is the distance between the plates, and e represents the permittivity of the material (permittivity is the measure of how easily a material can carry electric lines of force).

For the purposes of this section, we can define the area, A, as the trace length multiplied by the trace width. Wiring capacitance is determined as a capacitance per unit length for a given trace width and distance from the ground or power plane.

Let's examine a typical situation. For an eight layer PC board with 8 mil traces, and innermost layer ground/power planes, what is the capacitance per inch of trace on each of the signal layers?

Here are the terms we'll use in the equations to solve this problem and their values:

- trace width (w) = 8 mils (one mil equals 10^{-3} inch)
- trace length (l) = 1000 mils
- area (A) = w times l
- total board thickness (T) = 0.062 inch

- number of layers (N) = 8
- number of layers separating power and ground plane (n) = 1
- fringe effect and inter-trace stray capacitance adjustment factor (f) = 1.7
- permittivity of air (e) = $8.859*10^{-12}*($ coul^2 / (newton*m^2))
- relative permittivity of glass-epoxy dielectric (er) used in this example = 6

We start by determining the thickness of each dielectric layer, represented by t:

   t = T/(N - 1) = 8.857 mils

Next we need to determine the distance between the trace and ground/power plane, represented by d. This is found by the formula d = nt, which in this case makes for a simple calculation!

The capacitance as a function of the number of layers distance (Cd) is found by the formula:

   Cd = (ε * εr * A * f) / d

Using this formula,

   C(1 * d) = 2.073 pF (layer closest to ground/power plane)

   C(2 * d) = 1.037 pF (layer next closest to ground/power plane)

   C(3 * d) = 0.691 pF (layer farthest from ground/power plane)

To find the average capacitance per inch (Cavg), then

   Cavg = (C(1 * d) + C(2 * d) + C(3 * d))/ 3 = 1.267 pF

From this example, it is apparent that the stray wiring capacitance can vary significantly depending upon which layer of a multi-layer PC board a particular trace is located. Since a signal may travel on different layers between source and destination, exact values may be difficult to determine.

When performing a worst-case analysis of a given design, it is most effective to calculate the total load capacitance based on the sum of the loads' input capacitances, plus an estimate of the nominal wiring capacitance using 1 or 2 picofarads per inch of wiring using a rough guess for the length of the trace.

In a typical design, we might pick the diagonal distance from one corner of the board to the other, and multiply by 1 or 2 picofarads. If the total load capacitance is less than the driving device's specified test load capacitance, then the device will perform as specified. If not, or if it's very close, we might want to make a more accurate estimate, or avoid the problem by using a driving device that has a larger specified test load capacitance. Other alternatives include using two outputs *from the same chip* in parallel to double the drive capacity, or splitting the loads into two separate groups and driving them independently from two different sources.

As digital IC technology has improved, allowing signals to be processed at ever-increasing rates, the other non-ideal effects of the devices that could be ignored at lower speeds become more important. At very high speeds, these secondary effects become much more important. A wire ceases to be equivalent to a zero ohm connection with zero time delay. For the newer high-speed logic devices, the speed of the signal traveling down the wire, distributed resistance and inductance, as well as capacitance, may become very important. When the time it takes a signal to propagate down a wire are of the same order as the rise and fall time of the signal, it behaves as a *transmission line*, rather than an ideal wire. Transmission line effects are briefly described later in this chapter.

Fan-Out When CMOS Drives LSTTL

A common design problem involves the determination of how many LSTTL loads a CMOS output can drive. In this section, we will use the parameters below in an example to determine the number of LSTTL loads a CMOS gate can drive.

LSTTL gate DC Parameters:

Symbol	Parameter	min	typ	max	Units	Conditions
V_{IL}	Input Low voltage	-0.3		0.8	V	
V_{IH}	Input High voltage	2.4		Vcc+0.3	V	
I_{IL}	Input Low current		-120	-360	μA	
I_{IH}	Input High current		30	50	μA	
C_{IN}	Input Capacitance			10	pF	

Absolute Maximum Operating Conditions:

Symbol	Parameter	min	typ	max	Units	Conditions
V_{OL}	Output Low voltage		0.2	0.4	V	@ I_{OL} max
V_{OH}	Output High voltage	2.8	3.5		V	@ I_{OH} max
I_{OL}	Output Low current	3.2	8		mA	@ V_{OL} max
I_{OH}	Output High current	-600	-1000		μA	@ V_{OH} min

Note: Test conditions $R_L = 1K$, $C_L = 100\ pF$

CMOS gate DC Parameters:

Symbol	Parameter	min	typ	max	Units	Conditions
V_{IL}	Input Low voltage			2.0	V	
V_{IH}	Input High voltage	3.0			V	
I_I	Input leakage current			~ 0	μA	
C_{IN}	Input Capacitance			25	pF	

Absolute Maximum Operating Conditions:

Symbol	Parameter	min	typ	max	Units	Conditions
V_{OL}	Output Low voltage			0.4	V	@ I_{OL} max
V_{OH}	Output High voltage	4.5			V	@ I_{OH} max
I_{OL}	Output Low current	3.6			mA	@ V_{OL} max
I_{OH}	Output High current	600			μA	@ V_{OH} min

Note: Test conditions $R_L = 5K$, $C_L = 150\ pF$

For Logic One:

   CMOS I_{OH} = 600 microamperes (μA)
   LSTTL I_{IH} = 50 μA so 600μA/50μA = 12 loads

For Logic zero:

   CMOS I_{OL} = 3.6 milliamperes (mA)
   LSTTL I_{IL} = 360 μA so 3.6mA/360μA = 10 loads

Thus, considering the DC specifications only, the maximum number of loads driven is 10, since the zero state is the worst-case condition. The AC parameters would not be the limiting factor in this case, since the CMOS output is specified with a C_L of 150 pF, and each LS input is only 10 pF. Thus, 10 loads would present 100 pF plus stray wiring capacitance of less than 50 pF would present an AC load less than the 150 pF CMOS output load handling capability.

How many additional CMOS loads could be added? There are two levels of answer for this problem. First, from a DC point of view all the CMOS Iol output sink current is used up, so from this point of view, no loads could be added. However, there is negligible current in a CMOS input, so it is not the practical limit. In fact, the errors in the DC computations above are in excess of the amount required to drive a CMOS input, so in reality the DC current is not a problem. The real limitation is the capacitive loading. Even if you assume the loading from the TTL inputs and wiring can be ignored, the CMOS input capacitance will limit the loading. For the output to conform to the specs, the test load was specified as 150 pF (C_L). With ten LSTTL loads of 10 pF each, the C_L on the CMOS gate output would be 10 * 10 = 100 pF. Since the CMOS gate timing is specified at C_L =150 pF, there is only 150-100 = 50 pF left over to drive the additional CMOS loads. Since the CMOS Cin is 25 pF, the number of additional gates that can be driven is:

50 pF/25 pF = (remaining C_L) / (Cin of additional CMOS inputs) = 2

Practically speaking, the wiring capacitance on a PC board will generally be in the 2–3 pF per inch range, so allowing 25 pF for wiring capacitance would permit one CMOS load in addition to the 10 LSTTL loads from above.

What if the CMOS output were to drive only CMOS loads? The input capacitance of the CMOS gate is 25 pF, so even if *all* loads were CMOS, it can only drive C_L/Cin = 150 pF / 25 pF = 6 CMOS loads, and still meet its test condition limits. Since we must also allow for the wiring capacitance, we should limit this device to five loads, leaving 25 pF for the wiring capacitance. The additional load capacitance from more than five devices would likely result in timing performance that would be poorer than that specified in the data sheet. Excessive capacitance can also make ground bounce worse, which is the change in on-chip ground voltage due to rapid current spikes caused by charging load capacitance, developing a voltage across the lead inductance of the driving IC.

Transmission Line Effects

When using high-speed logic and the rise and fall times are of the same order as the propagation of the signal, transmission line effects become significant. When a signal transition propagates down a wire, it will be reflected back if the signal is not absorbed at the destination end. At lower speeds, the effect can be ignored, but with the fastest processors now in use, most designers will need to consider whether the effects will have a negative impact on their designs, and take appropriate action if necessary.

There are several characteristics of digital transmission lines that must be addressed, including the following:

- signal transition time vs. clock rate
- mutual inductance and capacitance (crosstalk)
- physical layout effects
- impedance estimates
- strip line vs. micro strip
- effects of unmatched impedances
- termination and other alternatives
- series termination vs. parallel termination
- DC vs. AC termination techniques

The techniques for high speed design are beyond the scope of this text, and are covered in detail in an excellent text on the subject, *High Speed Digital Design, a Handbook of Black Magic*, by Howard W. Johnson and Martin Graham. In contrast with the subtitle, this subject is easily understood by applying some very basic physics.

A transmission line is a conductor long enough so that the signal at the far end of the line is significantly different from the near end, due to the time it takes the signal to propagate from one end to the other.

In this book, we will assume that the interconnections between the devices are *not* long enough to require transmission line analysis. In order to verify that this is the case we can use a simple estimate. The rough estimate we will make is based on the idea that a wire does not have to be analyzed as a transmission line if the signal takes longer to rise or fall than it takes to get from one end of the wire to another. In other words, if the signal doesn't have to travel too far, both ends of the wire are at approximately the same voltage. In order to come up with a numerical value to determine if a signal must be treated as a transmission line, there is a simple calculation that can be used, shown below.

$l = T_r / D$, where
$\quad$ l = length of rising or falling edge in inches (in)
$\quad T_r$ = rise time in picoseconds (pS)
$\quad$ D = delay in picoseconds per inch (pS/in)

For traces on a standard printed circuit board, the value for D will be in the range of 100 to 200 pS/in. Depending upon how much distortion you're willing to live with, the critical trace length will be between one-sixth and one-quarter of the length of a trace corresponding to the signal's transition. For a trace that is shorter than one-sixth the length of the signal's rising or falling edge, the circuit seldom needs to be considered to be a transmission line. Traces that are much longer than one-quarter the length of the fastest edge will start to behave as transmission lines, exhibiting reflections of the signal when the transition gets to the far end of the trace and is reflected back to the near end. Once the trace is about half of the length it takes for a logic transition to propagate, the problems become quite pronounced.

Let's look at an example. A logic device on a standard glass-epoxy printed circuit board has a 2 nS rise time.

This signal has a rising edge that is:

(2 nS)/(150 pS/in) = ~13 inches long.

That means a trace that is one-sixth that length, or about two inches or less, does not have to be considered as a transmission line. If the trace is much longer than two inches, it will begin to show significant distortions on the rising and falling edges due to the fact that there is a different signal voltage at each end of the trace at the same instant, resulting in reflections of the signal from the ends of the trace.

This is one of the most important reasons for using logic that is fast enough, and not too much faster than required to meet the timing requirements. While it might seem tempting to buy the fastest device available to reduce the delays in a device which does not meet the timing requirements, doing so can result in a lot more difficult problems to solve!

Ground Bounce

Another effect of high-speed signal transitions is called *ground bounce*. Ground bounce occurs when a large peak current flows through the ground pin of a chip when one or more logic outputs change state and discharge their load capacitances through the chip's ground pin. While the parasitic inductance of the ground pin may not seem very significant, in the nanohenry (10^{-9} H) range, fast transients can cause large voltages to appear across the ground pin. This occurs most often when multiple bus signal outputs from one chip change state at the same time. The rapid, parallel current pulses which result from charging or discharging stray bus capacitance must be carried through the ground or power pins, which have inductance.

The voltage across an inductor is equal to the inductance times the rate of change of current through the inductor, or:

$V = L * di/dt$, where
    V = instantaneous voltage across the inductor (volts)
    L = Inductance (henry)
    di/dt = Rate of change of current (amperes/sec)
    and current i = Q/t (amperes = coulombs per second)

The charge on a capacitor is Q = CV (coulombs = farads * volts)

$V = L * C * (\text{delta } V) / (\text{delta } t)^2$ *approximately,* or
$V = L * C * (V_{oh} - V_{ol}) / (T_r)^2$ using the output voltage and rise time

Because of the high-speed (nS) and large (amperes) peak currents, even the small nanohenry inductance can induce a voltage transient on the order of volts. (The instantaneous voltage across an inductor is $V = L * di/dt$.) For typical high speed signals nanohenries*amperes/nanoseconds = volts! This effect is minimized by the use of minimum circuit interconnect trace lengths, wider ground traces, power and ground planes, and small, surface mounted IC packages that have very short leads.

For example, a CMOS output driving a 100 pF load with a rise time of 2 nS would induce a voltage across a typical 1 nH inductance of the chip's ground lead:

$V = 1 \text{ nH} * 100 \text{ pF} * (4.5 - 0.5 \text{ V}) / (2 \text{ nS})^2 = 0.1 \text{ V}$

While a voltage of 0.1 volt or 100 millivolts may not seem like much, remember that a part with many outputs, such as a processor, will sometimes switch many outputs at the same time, and *the current that flows through those pins all has to flow through a single ground pin.* An 8-bit output will cause 0.8 volt pulse or ground bounce. If the processor drives an 8-bit data bus and a 16-bit address bus low at the same time, this would result in a 2.4 volt bounce! The ground bounce voltage across the ground lead inductance results in a different ground voltage reference for the chip while the chip's ground is bouncing. Needless to say, this ground bounce can cause a logic level to change during the brief pulse, which can cause trouble with circuits, such as clock signals, which are edge sensitive. This is why high-speed logic devices may have multiple, short ground pins, and may only be available in small, surface mounted packages. To make things even worse, if two devices overlap slightly in time driving the bus, very large current transients may briefly generate even larger currents that in turn generate larger ground bounce pulses. This can disturb several chips on the board at the same time.

The power supply leads are also subject to bounce for exactly the same reasons, and even though the power supply is not used as a logic voltage reference, the resulting drop in the local power supply voltage to the chip can result in errors.

While exact ground lead inductances may prove difficult or impossible to measure, there is always some inductance in the ground lead, and the longer the lead, the greater the inductance. The example above illustrates another reason why it makes sense to avoid logic that is faster then necessary, and to use very short ground and power wires. In fact, high speed PC boards should use separate inner layers of a multi-layer board to provide large ground and power planes, allowing the chips' power and ground leads to be connected using very short wires.

The magnitude of the bounce depends upon the number and direction of logic transitions, so the noise is also data dependent! This is an apparently intermittent hardware design fault with symptoms that act like a software bug, since it may only happen at certain points in executing a program, with certain data values.

The example also shows why it is so important to maintain sufficient tolerance to noise in the logic. This noise tolerance is referred to as noise margin, which is covered in the next section. Noise margin analysis is especially important in a high-speed logic design, to prevent transient logic errors, which are extremely difficult to track down. This is another example of how a proper analysis and worst-case design can save a lot of time and money while delivering much higher quality and ultimately reliability. In the next section, the noise margin analysis process is described in detail.

Logic Family IC Characteristics and Interfacing

The three most common logic families are:

- **TTL**: transistor-transistor logic (also known as bipolar logic)
- **NMOS**: n-channel metal oxide semiconductor field effect transistor logic
- **CMOS**: complementary (n- and p- channel) MOS logic

All three logic families have versions with TTL compatible inputs, once the most common type, followed by later NMOS and CMOS. Because of its lower power density and relatively high circuit density however, CMOS has become the most common form of logic, particularly in high density and low power battery operated systems. TTL logic uses bipolar transistors requiring input drive currents on the order of hundreds of microamperes to a few milliamperes, depending on the version. Input voltage ranges for TTL level compatible logic are generally 0 to 0.8 volts for logic zero and 2.4 to 5 volts for logic one. Output voltages are from 0 to 0.4 volts for logic zero and 2.8 to 5 volts for logic one. The 0.4 volt difference is called the *noise margin* voltage because additive noise at or below this level will not change zeros to ones or vice-versa. The *logic threshold voltage* (V_T) or "0/1 decision point" for TTL logic is typically around 1.5 volts. It may range anywhere between 0.8 and 2.0 volts depending upon supply voltage, temperature, and varies from one device to another. For TTL circuits, the noise margin is at least 0.4 volts. Figure 3-10 shows the concepts of noise margin and logic threshold voltages.

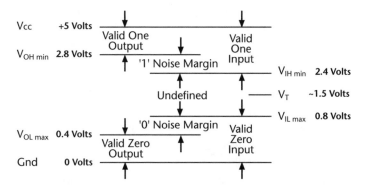

Figure 3-10: Typical TTL logic voltages and noise margin.

Interconnecting different logic families, such as CMOS and TTL, requires the designer to assure the compatibility of the logic signal voltage levels, and adapt the circuit as necessary to maintain appropriate noise margins. The equivalent resistance or impedance of the signal network also has an impact on the noise in a specific circuit. High impedance inputs are more prone to noise than low impedance inputs. The interface design process is illustrated by an example at the end of this chapter.

TTL logic is capable of sinking high currents and is used for driving very fast, large, heavily loaded buses. Both active and passive pull-up output devices are used with TTL. The active pull up, referred to as a *totem-pole output* uses one transistor to source current and one to sink it. The passive pull-up uses a transistor to sink current, and a resistor connected to V+ as a current source. If a pull up resistor is not connected to the gate's output pin, and the collector is connected only to the output pin, it is referred to as an *open collector output* In both cases, the output current sinking capabilities are greater than current source capacity. Many devices can sink a few milliamperes, but can only source hundreds of picoamperes. Figure 3-11 shows both totem pole and open collector outputs.

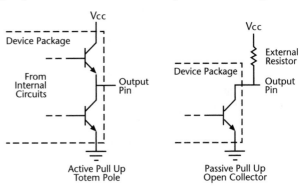

Figure 3-11: TTL outputs, totem pole and open collector.

TTL and CMOS logic are available in several versions, each identified by a distinctive prefix in the part number. Some of the more common versions and their prefixes are:

74xx:	standard TTL
74LSxx:	low power Schottky clamped TTL
74ALSxx:	advanced LS TTL
74Fxx:	(fast) high speed TTL
74HCxx:	high speed CMOS with CMOS compatible inputs (Vt = ~Vcc/2)
74HCTxx:	high speed CMOS with TTL compatible inputs (Vt = ~1.5V)
74FCTxx:	high speed CMOS with TTL compatible inputs (Vt = ~1.5V)
74ACTxx:	advanced high speed CMOS with TTL compatible inputs
74BCTxx:	very high speed CMOS/Bipolar with TTL compatible inputs

Schottky logic (74ALSxx 74LSxx and 74Sxx) incorporates a low V$_f$ (forward voltage drop) Schottky diode across the collector-base junction of a transistor to prevent it from saturating. This increases the speed for turning the transistor off. TTL is generally used where low cost, output drive, and high speed are important, and there is no objection to the relatively high power consumption and resulting heat.

NMOS logic was used for moderate complexity logic ICs such as more mature microprocessors. Most NMOS logic ICs have TTL compatible voltage specs and operate at a lower power and speed than TTL. The power consumed by NMOS lies between TTL and CMOS, as does its speed. The input current is nearly zero since the MOSFETs have extremely high input resistance. Unfortunately, they do have fairly large input capacitance, limiting the circuit speed. The output configurations are similar to TTL except the transistors are n-channel field effect transistors (FETs) rather than bipolar NPN. Both active totem pole and passive (open drain) outputs are used in microprocessor and microcontrollers. Because of the constant operating current drain, these devices tend to be limited in size and complexity.

CMOS logic has a significant advantage since it does not use any significant amount of power when it is static (not changing state). Most of the power used in an operating device is due to the charge and discharge of internal capacitance and the current transient when both N and P devices are partially on. As a result, power consumption is a function of clock rate for CMOS devices. Some processors are even designed to take advantage of this fact by incorporating "sleep" or low power modes stopping some or all of the clock operations when nothing important is going on. This is frequently required for battery-operated systems to maintain a reasonable battery life. Another advantage is the standard CMOS logic threshold is one half the supply voltage, and the output voltages tend to be very close to Vcc and ground voltage, resulting in higher noise margins than those of TTL devices. This is particularly important for CMOS devices that operate at reduced power supply voltage. CMOS devices are available which operate at 3 volts or less.

Because CMOS logic is inherently symmetrical, the rise and fall times tend to be nearly equal. The symmetry also results in equal source and sink capabilities. The inherent increase in noise margin makes CMOS less susceptible to noise than TTL and NMOS. Figure 3-12 illustrates this. CMOS devices operating at voltages other than 5 volts, such as 3.3 volts, will have a threshold voltage corresponding to Vcc/2. Some versions of CMOS logic operate with a reduced noise margin in order to have TTL compatible input voltages. This is accomplished by artificially lowering the input threshold voltage to 1.5 volts, the same as used for TTL. These TTL input threshold compatible circuits have a T in their number (74HCT, 74BCT, etc.) indicating they have TTL compatible inputs. A series of high-speed logic compatible with the TTL logic family in function and input voltage is the 74HCTxx (High speed CMOS TTL compatible) series. The advantage of the 'T' series CMOS devices is they can be driven directly by devices having TTL output voltage levels. The 'T' series of CMOS devices has the disadvantage that the noise margin is less than it is for true CMOS compatible inputs due to the shifted threshold voltage. The 74HCxx series is pure CMOS with a threshold voltage of one-half the supply voltage (2.5 volts for a 5 Vcc) and correspondingly higher noise margins. As a result, a standard TTL output V_{OHmin} of 2.8 volts is not enough to guarantee a logic one value for a 74HCxx gate input.

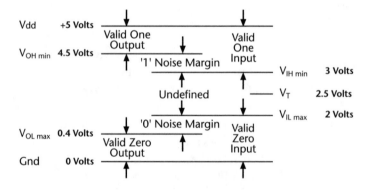

Figure 3-12: Typical CMOS logic voltages and noise margin.

Interfacing TTL Compatible Signals to 5 Volt CMOS

Interfacing a CMOS output to a TTL input is a direct connection, as long as the CMOS output is capable of sinking the TTL device's input low current. Interfacing a TTL output to a standard CMOS input requires the use of at least a pull up resistor. A resistor on the TTL output to Vcc will ensure the output voltage is pulled high enough to guarantee the logic one output signal is interpreted as a logic one by the CMOS input. Another useful technique when using 5 volt logic to drive CMOS circuits, is to use a higher voltage open collector or open drain output with a pull up resistor connected to the higher supply voltage. This level shifting technique can also be used for driving other high voltage circuits such as high voltage outputs. In either case, the objective is to guarantee there is sufficient noise margin to guarantee a valid logic one when the TTL compatible output drives a CMOS input.

It is important to note that when a TTL output is pulled above its normal output high voltage, it will not source any significant current. This is because the TTL output source is equivalent to a high resistance in series with a voltage source that is effectively limited to around 3 volts, due to internal design constraints. As the output voltage increases until it equals the internal voltage, the output can no longer source any current. When the voltage is increased beyond the internal circuitry (up to a limit of Vcc), the internal circuitry is equivalent to a reverse biased diode, so only leakage currents in the sub-microampere range will flow into the output device. As a result, the effect of a TTL output on external circuits is negligible when the pin is pulled high by an external resistor.

Also, a 5 volt TTL compatible output is often compatible with a 3 volt CMOS device input, since the CMOS threshold (Vcc/2 = 1.5 volt) is the same as a 5 volt TTL gate (TTL Vt = 1.5 volt). Most of the 3 volt CMOS devices are designed to withstand a 5 volt input signal, so it is often possible to interface 5 volt TTL outputs directly to 3 volt CMOS inputs. However, if the 3 volt CMOS inputs are not designed to handle 5 volt inputs, the CMOS device could be destroyed with an input signal greater than 3 volt, so it is important to verify this. A 3 volt CMOS device output will be close to 3 volt, so it can drive a 5 volt TTL compatible input directly.

A 3 volt CMOS output would probably be marginal driving a 5 volt CMOS input (Vt = Vcc/2 = 2.5 volt), leaving less than 0.5 volts of noise margin. Since the 3 volt CMOS output generally cannot withstand a pull-up resistor to 5 volts, it is necessary to add a level shifting IC to convert 3 volt logic levels to 5 volt.

Level shifters are available for converting logic levels from one family to another, including 3 volts to and from 5 volt, or 5 volt TTL to +/- V ECL (*emitter-coupled logic*), and 5 volt levels to +/-12 volt RS-232 signals. There are also special ICs for driving output loads requiring either a high voltage or high current output, such as a light, motor or relay. Most microcontrollers have very weak output drive capability, so external driver ICs may be necessary. These would typically be needed to drive LEDs, a vacuum fluorescent display, or a motor. Solid-state relays even allow large AC loads to be controlled by a micro. Likewise, there are other devices (i.e., optical isolators), allowing high voltages (like 110volt AC inputs) to be safely converted to logic levels for input to a microcontroller. Devices that use potentially hazardous high voltages should be isolated from the rest of the circuitry for reasons of safety. While it may be possible to connect such devices directly to our circuits, they would allow us to come into contact with potentially fatal voltages. Unfortunately, the standard 50 or 60 cycle AC power supply used almost everywhere has the unfortunate characteristic that it is very nearly the optimal voltage to guarantee that a human heart will stop functioning due to muscle fibrillation. Customer death by electrocution is sure to result in the next of kin hiring an attorney to relieve you of all your assets. . . . unless, of course, they're *your* next of kin! There are many isolation devices available, most of which use the same basic approach.

The isolation can be accomplished using optical or magnetic means, which can provide a barrier to transient voltages that can be on the order of thousands of volts. The barrier is transparent, and so allows light to pass, but is made of a good insulator to prevent electrical current from flowing across the boundary. Figure 3-13 shows a simple optical isolation circuit.

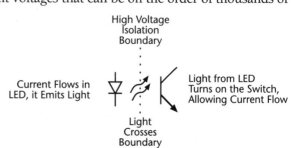

Figure 3-13: Optical isolation allows connection to hazardous voltages.

This isolation approach can be used to input high voltages to a microcontroller safely by connecting the LED to a high voltage source in series with a resistor and protective diode to limit the LED's current and prevent the LED from being exposed to the potentially destructive reverse voltage. The output transistor will then be turned on whenever the LED is turned on by one half of the AC power cycle. This is useful for time of day clock functions, since the AC power mains frequency is maintained very accurately by the power utilities over a period of time. The output switch can be connected to the processor counter or interrupt input, allowing the processor to keep track of time and synchronize its operation with the AC power cycle.

High voltage outputs can also be controlled safely by using the micro's output to turn on the LED that turns the output switch on. In this case, another type of switch such as an *SCR* (silicon-controlled rectifier) or *TRIAC* (an AC version of the SCR) is used rather than a transistor. SCR and TRIAC switches can be obtained to handle relatively large AC loads, such as lamps, and motors. These devices are often referred to as *solid-state relays* (SSR), since they are equivalent to an electromechanical relay, except that they are implemented with solid-state semiconductor devices instead of using a coil to move a switch. Both isolated inputs and outputs are available in complete modules that have all the necessary circuits to monitor and control high voltage and power devices, using optical isolation for safety. They have microcontroller compatible I/O on one side that is optically isolated from the high power outputs on the other side.

Very often, even when safety is not an issue, microcontroller chips simply cannot handle the voltages or currents required to interface with other devices. In some cases it is required when connecting one logic family to another, incompatible family, such as emitter-coupled logic (ECL) levels or RS-232 interfaces utilizing negative voltages.

Sometimes, a plain, old-fashioned electromechanical relay is a better solution, since relays usually have contact resistances that are far lower than can be found in a semiconductor switch. In some cases, a simple transistor or MOSFET switch can be used to control a load operating at voltages which are greater than the logic supply, such as motors, solenoid actuators, and relays which may require 12 or more volts to operate.

The circuitry required to interface between logic levels and high-level circuits is described in detail elsewhere, including an excellent book titled *The Art of Electronics*, by Horowitz and Hill. If you don't already have this book—and you have to do much electronic design or interfacing—you should definitely obtain a copy of this very handy book.

The real world is an analog place, and interfacing between the discrete, digital world of computers and the real world demands significant attention. The interface between low level analog signals and logic is handled in another chapter of this book.

At this point, it is time to look at some simple examples, so we can see exactly how a worst-case analysis should be performed. The next section illustrates part of the worst-case analysis for a real laboratory instrument that is still used in the healthcare industry. This product's poor reliability was seriously inconvenient for the medical staff and patients who depend upon it, and if it had lead to an incorrect diagnosis, a truly fatal error! It is in these types of applications that worst-case design is most important, and the cost of unreliable hardware in the field almost always greatly exceeds the cost of avoiding the problem by using proper design and analysis techniques. Now let's turn our attention to the analysis of the worst-case noise margin for an 8051 based design example.

Design Example: Noise Margin Analysis Spreadsheet

The following spreadsheet shows the results of a noise margin on a design that was already in production at the time of the analysis. The product's users had complained about intermittent glitches, and the author was consulted to determine the source of the problem. After a quick look at a few of the noise margin values, it became obvious that there were deficiencies in the design in that area. A portion of the spreadsheet used in that analysis is shown in Table 3-1, with problems shown in ***bold italic underline*** font.

The first column of Table 3-1 is the signal name, followed by the pin number and chip which is the source of the signal, followed by the source's worst-case output voltages, Volmax and Vohmin. The next columns list the loads on the signals and their respective worst-case input voltages Vilmax and Vihmin. The noise margins are shown in the last two columns, Vil - Vol for the logic zero case, and Voh - Vih for the logic one case. As can be seen, the logic zero noise margins are all probably acceptable, as the lowest value is 0.3 volts. The logic one noise margin is zero or negative for most of the devices listed, which is completely unacceptable. Any noise on the power supply, ground or the signal lines themselves can easily cause a logic input to interpret the wrong logic state, causing an error. An interesting thing to observe is that none of them were very far out of spec, and the instrument worked perfectly most of the time. These problems can be virtually impossible to find in the field. Hooking up a test instrument like a scope or logic analyzer to the problem signals often makes the problem go away, due to changing the ground currents and impedances of the circuit. The specs that cause the problem in this case are the high Vih specs of the loads, especially the SRAM chip. The example design in the sheet above represents a relatively common problem with devices that are advertised as "compatible" with other logic families. The solution to the prob-

8051 Noise Margin Analysis - Sample

OUTPUT					INPUT				Noise	Margin
			Vol	*Voh*			*Vil*	*Vih*	*logic*	*logic*
Signal	*Pin(s)*	*Source*	*max*	*min*	*Load(s)*	*Signal*	*max*	*min*	*zero*	*one*
PSEN/	29	8051	0.40	2.00	EPROM	OE/	0.80	2.00	0.40	*0.00*
RD/	17	8051	0.40	2.00	SRAM	OE/	0.80	2.20	0.40	*-0.20*
(P3.7)			0.40	2.00	82C55	RD/	0.80	2.00	0.40	*0.00*
WR/	16	8051	0.40	2.00	SRAM	WR/	0.80	2.20	0.40	*-0.20*
(P3.6)			0.40	2.00	82C55	WR/	0.80	2.00	0.40	*0.00*
A15 (P2.7)	28	8051	0.40	2.00	74LS138	A	0.80	2.00	0.40	*0.00*
A8..14	21-27	8051	0.40	2.00	SRAM	A8..14	0.80	2.20	0.40	*-0.20*
(P2.0-P2.6)			0.40	2.00	EPROM	A8..14	0.80	2.00	0.40	*0.00*
			0.40	2.00	GAL	A8..14	0.80	2.00	0.40	*0.00*
ALE	30	8051	0.40	2.00	74LS373	LE	0.80	2.00	0.40	*0.00*
AD0..7	39-32	8051	0.40	2.00	74LS373	A0..7	0.80	2.00	0.40	*0.00*
(P0.0-P0.7)			0.40	2.00	SRAM	D0..7	0.80	2.20	0.40	*-0.20*
			0.40	2.00	82C55	D0..7	0.80	2.00	0.40	*0.00*
		SRAM	0.40	2.20	8051	D0..7	0.80	2.40	0.40	*-0.20*
		EPROM	0.45	2.40	8051	D0..7	0.80	2.40	0.35	*0.00*
		82C55	0.40	3.50	8051	D0..7	0.80	2.40	0.40	1.10
RAM Enable	16V8		0.50	2.40	SRAM	/CE	0.80	2.20	0.30	0.20
EPROM En.	16V8		0.50	2.40	EPROM	/CE	0.80	2.00	0.30	0.40

Table 3-1

lem is very simple and inexpensive: the addition of pull-up resistors to the
signals that have zero or negative noise margin in the logic one state. This also
impacts the output low current that must be handled by the signal source
chip outputs, so it must be taken into account in the load analysis and pull up
resistors should be chosen accordingly.

It is important to note that there are four sources listed for AD0..7, since there are four devices that drive the data bus. Only the data paths that are used need to be evaluated vs. loading analysis, where unused paths load the bus. The load analysis for another similar design is shown in Table 3-2, which tabulates the capabilities of the various driving devices, and the loads that are presented to them. The first three columns (signal, pin and source) identify the signal source, the next three (IOL, IOH and CL), list the corresponding source's output drive current and capacitive load values. The next two columns (load, and signal) identify the load's signal names. The Qty column is the number of loads in the case of multiple signals connected to the same output, or the number of inches of wire in the case of the wire capacitance. The next three columns (IIL, IIH, and Cin) define the load characteristic of a single input's input current and input capacitance. For the interconnect wiring, Cin is the estimated stray wiring capacitance per inch of the printed circuit trace. The last three columns show the extended totals and grand totals for each signal, followed by the design margin, which should be a positive number. In this case there is only one problem, due to excessive capacitive loading of the SRAM when it drives the data bus, AD0..7.

The output capacitive load specs are usually found as notes within the AC section of the chip specification listing the various timing parameters. This is because the capacitive loading affects the rise and fall time of the signal, so the capacitance value is really used as a test condition for the timing measurements. Input capacitance may be difficult to find in the specification sheet, it may be in a different "family" specification sheet or handbook, or may not be specified at all. When it is not specified, a reasonable estimate can be made by substituting values for similar parts in the same type of package.

The SRAM output is specified with a Cload value of 50 pF, which is relatively low value. By using a very low load capacitance, the SRAM's timing specs look good due to shorter than normal rise and fall times, since the chip is not driving a realistic load. This is a good example of a manufacturer's "specsmanship." They are intentionally playing games with the test conditions to make their device appear to be better than it is. That way when someone looks at their timing specs, the shorter rise and fall times make their chip appear to be faster than another equivalent chip that is specified with a larger capacitive load value, when the chips are actually identical. Unfortunately, this practice is all too common, so that the designer must view the claims on the cover of a data sheet very critically. If it looks to good to be true, then it probably is!

Table 3-2

Signal	Pin#	Source	uA IOL	uA IOH	pF CL	Load	Signal	Qty	uA IIL	uA IIH	pF Cin	uA IIL	uA IIH	pF Cin
PSEN/	29	8051	3200	-60	100	EPROM	OE/	1	-1	1	12	-1	1	12
						wire cap		2			2			4
											Total	-1	1	16
											Margin	3199	59	84
RD/	17	8051	1600	-60	80	SRAM	OE/	1	-1	1	7	-1	1	7
(P3.7)						82C55	RD/	1	-1	1	10	-1	1	10
						wire cap		3			2			6
											Total	-2	2	23
											Margin	1598	58	57
WR/	16	8051	1600	-60	80	SRAM	WR/	1	-1	1	7	-1	1	7
(P3.6)						82C55	WR/	1	-1	1	10	-1	1	10
						wire cap		3			2			6
											Total	-2	2	23
											Margin	1598	58	57
A15	28	8051	1600	-60	80	74LS138	A	1	-200	20	10	-200	20	10
(P2.7)						wire cap		2			2			4
											Total	-200	20	14
											Margin	1400	40	66
A8..14	21-7	8051	1600	-60	80	SRAM	A8..14	1	-1	1	7	-1	1	7
(P2.0-P2.6)						EPROM	A8..14	1	-1	1	12	-1	1	12
						wire cap		3			2			6
											Total	-2	2	25
											Margin	1598	58	55
ALE	30	8051	3200	-60	100	74LS373	LE	1	-400	20	10	-400	20	10
						wire cap		2			2			4
											Total	-400	20	14
											Margin	2800	40	86
AD0..7	39-2	8051	3200	-800	100	74LS373	A0..7	1	-400	20	10	-400	20	10
(P0.0-P0.7)						SRAM	D0..7	1	-1	1	7	-1	1	7
						EPROM	D0..7	1	-1	1	12	-1	1	12
						82C55	D0..7	1	-10	10	20	-10	10	20
						wire cap		5			2			10
											Total	-412	32	59
											Margin	2788	768	41
		SRAM	1600	-600	50	74LS373	A0..7	1	-400	20	10	-400	20	10
						8051	D0..7	1	-1	1	20	-1	1	20
						EPROM	D0..7	1	-1	1	12	-1	1	12
						82C55	D0..7	1	-10	10	20	-10	10	20
						wire cap		5			2			10
											Total	-412	32	72
											Margin	1188	568	-22
		EPROM	1600	-600	100	74LS373	A0..7	1	-400	20	10	-400	20	10
						SRAM	D0..7	1	-1	1	7	-1	1	7
						8051	D0..7	1	-1	1	12	-1	1	12
						82C55	D0..7	1	-10	10	20	-10	10	20
						wire cap		5			2			10
											Total	-412	32	59
											Margin	1188	568	41
		82C55	1600	-60	80	74LS373	A0..7	1	-400	20	10	-400	20	10
						8051	D0..7	1	-1	1	20	-1	1	20
						EPROM	D0..7	1	-1	1	12	-1	1	12
						SRAM	D0..7	1	-1	1	7	-1	1	7
						wire cap		5			2			10
											Total	-403	23	59
											Margin	1197	37	21

When an output like this is operated with actual capacitive load greater than the test conditions, the related timing specs for the device must be de-rated, due to the degraded rise and fall times that will occur. As long as the load capacitance is no more than twice the spec value, this will be sufficient. The excess C load will increase the stress on the driver. If the overload is much greater than two times normal, the device can be overstressed due to the relatively large currents that will flow into the load capacitance on transitions when the C is charged and discharged through the driving output. As long as the output is not overloaded too much, the resulting increase in the rise/fall time can be estimated, resulting in a de-rated timing spec. All we have to do is calculate the additional rise time and add that to the timing values specified in the data sheet. In order to do that, we need to evaluate the output circuit's performance. This can be accomplished by noting that the output current drives the load capacitance from a logic low to high or vice versa. For our purposes, we will assume that the interconnect does not behave like a transmission line, which is most often the case for garden variety microcontroller components. If the chips used have a fast rise time and trace length greater than about one-sixth the edge length of the pulse, then it is necessary to analyze the circuit as a transmission line. In this case we will look at the simpler problem.

By assuming a constant current charging the capacitance, the voltage will ramp linearly from one logic level to the other. To make a rough estimate, we can use the source's output current and load capacitance to determine the signal slew rate, and the difference between the high and low logic levels to determine the delay. Figure 3-14 illustrates this.

Let's next look at a simple example showing how to de-rate the timing based on the approximation technique just described.

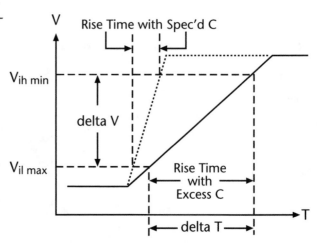

Figure 3-14: Derating delay for excess CL.

First we make the assumption that the signal timing measurements in the data sheet are made under the specified test conditions, usually with the output loaded by R_L and C_L in parallel to ground. The output delay specifications in the data sheet include the internal delay as well as the rise time. The output drive current charges C_L within the specified time. The circuit can be divided into two parts: the specified load, and the additional output current available to drive the excess load C. So the additional delay (delta T) we are looking for depends upon the leftover drive current (delta I) which is available to charge the excess load capacitance (delta C). The equation for this is:

Delta T = (delta V * delta C) / (delta I)

Let's look at a typical example. An SRAM is specified with a 50 nS access time, but the outputs are overloaded with respect to the C_L spec in the data sheet. What access time spec should be used for the actual conditions specified below?

The output is specified to drive C_L = 50 pF, but the actual load is 100 pF.

The output is specified to drive 20 mA into the load, but the load is only 10 mA.

The driven device has input voltage specs Vilmax = 0.4 V, Vihmin = 3.4 V.

Spec values:	Actual Values:	Difference:
C_L = 50 pF	100 pF	50 pF = delta C
Io = 20 mA	10 mA	10 mA = delta I

Voltage: Vih - Vil = 3.4 - 0.4 = 3 V = delta V

Delta T = (delta V * delta C) / (delta I)

Delta T = (3 V * 50 pF) / (10 mA) = 15 nS

So in this case 15 nS should be added to all the output delay specs for the driving device. The access time used should be:

Taa(actual) = Taa(spec) + (delta T) = 50 nS + 15 nS = 65 nS

Since the output current from most devices is larger at the beginning of the transition and smaller near the end of the transition, the approximation is only a rough guide. Also, the delta V calculation is conservative, since the input threshold voltage is typically half way between the Vih and Vil values.

So, the estimate as shown will usually be conservative compared to actual performance. All of the above must be used with caution, and is only an approximation of the additional delay caused by excess C_L, so it is wise to allow additional margin in the timing for any de-rated specs.

Here's another typical example. An LSTTL gate is to be used to drive one LSTTL load and a CMOS processor clock input, as shown in Figure 3-15. An interface must be made which will guarantee the CMOS input voltage requirement will be met with the same noise margin as a standard LSTTL input. The LSTTL and CMOS gates have the specs as defined below:

LSTTL Gate DC Parameters

Symbol	Parameter	min	typ	max	Units	Conditions
V_{IL}	Input Low voltage	-0.3		0.8	V	
V_{IH}	Input High voltage	2.4		Vcc+0.3	V	
I_{IL}	Input Low current		-120	-360	μA	
I_{IH}	Input High current		30	60	μA	

Absolute Maximum Operating Condition:

Symbol	Parameter	min	typ	max	Units	Conditions
V_{OL}	Output Low voltage		0.2	0.4	V	@ I_{OL} max
V_{OH}	Output High voltage	2.8	3.5		V	@ I_{OH} max
I_{OL}	Output Low current	3.2	8		mA	@ V_{OL} max
I_{OH}	Output High current	-600	-1000		μA	@ V_{OH} min

Note: Test conditions $R_L = 1K$, $C_L = 100\ pF$

CMOS Gate DC Parameters

Symbol	Parameter	min	typ	max	Units	Conditions
V_{IL}	Input Low voltage			2.0	V	
V_{IH}	Input High voltage	3.0			V	
I_I	Input leakage current			<1	μA	

Absolute Maximum Operating Conditions:

Symbol	Parameter	min	typ	max	Units	Conditions
V_{OL}	Output Low voltage			0.4	V	@ I_{OL} max
V_{OH}	Output High voltage	4.5			V	@ I_{OH} max
I_{OL}	Output Low current	3.2			mA	@ V_{OL} max
I_{OH}	Output High current	600			μA	@ V_{OH} min
C_{in}	Input Capacitance			20	pF	

Note: Test conditions $R_L = 5K$, $C_L = 150\ pF$

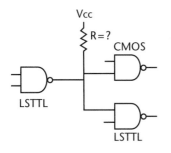

Figure 3-15: TTL to CMOS interface example.

Here is how we would determine the answer. Since the LSTTL V_{OL} is 0.4 volts and the CMOS V_{IL} is 2.0 volts, the CMOS input low voltage is compatible with the LSTTL low output voltage. However, the LSTTL output high voltage of V_{OH} = 2.8 volts is not sufficient to meet the CMOS input high V_{IHmin} = 3.0 volts. A pull-up resistor is required to allow the LSTTL output to go to a higher voltage, $V_{IH} + V_{noise\ margin}$ = 3.0 + 0.4 = 3.4 volts. There is no exact solution, but the range of resistors meeting the requirements can be determined.

The lowest resistor value that will work is the value which will source enough current so the LSTTL output is just able to sink the resistor current plus the additional LSTTL load when the signal is low and still meets the maximum output low voltage specification. There is negligible DC current flowing from the CMOS input. The voltage across the resistor is $Vcc - V_{OL\ max.}$ for the LSTTL input, or 5 – 0.4 = 4.6 volts. The current required is $I = I_{ILmax} + I_{RPU}$ where I_{ILmax} is the current coming from the LSTTL input load and I_{RPU} is the current flowing through the pull up resistor. The current the LSTTL output must sink is the sum of the I_{IL} of the LSTTL load and the current through the pull up resistor.

The equation is:

$$I_{OLmin} >= I_{ILmax} + I_{RPU} = 360\ \mu A + (Vcc - V_{OL\ max}) / R_{min}$$

Solving for R_{min} :

$$R_{min} > = (5 - 0.4\ volts) / (3.2\ mA - 360\ \mu A) = 4.6\ V / 2.84\ mA = 1.62\ kilohms$$

R_{min} is 1.62 Kilohms

This value is also greater than specified as a test load of 1 kilohm.

The maximum acceptable value, R_{max}, is determined by the minimum output high voltage that will guarantee a CMOS high input plus noise margin. The resistor must be able to supply the LSTTL maximum input high current and not have too large a voltage drop across it. This will determine the upper limit for the resistor value.

Specifically, the resistor voltage is:

$$Vcc - (CMOS\ V_{IH\ min} + V_{noise\ margin}) = 5 - (3.0 + 0.4) = 1.6\ volts$$

This voltage is maintained while sourcing the LSTTL $I_{IH\,max}$ of 60 μA.

Solving for R_{max} :

   R_{max} <= 1.6 V / 60 μA = 26.7 kilohms maximum

Thus, the acceptable range for the pull up resistor is

   1.62 kilohms <= R_{PU} <= 26.7 kilohms

An acceptable standard value such as 10 kilohms would be appropriate.

Another limit relates to the rise time of the signal under load, due to the R-C time constant of the pull-up resistor charging the load capacitance, C_L. From the example above, let's see what the effect of this time constant is on the selection of the resistor value.

The maximum R value can be approximated by the equation:

   $R = T / C_L$ where T is the rise time and C_L is the total load capacitance

Ignoring the Ioh current of the LSTTL driver, if the circuit above had an allowable rise time T = 50 nS and C_L = 20 pF, then the maximum R value would be:

   R_{max} = 50 nS / 20 pF = 2.5 kilohms maximum to maintain the 50 nS rise time.

So a better choice might be a standard 2.2 kilohm pull-up resistor. Since the driver will supply some current to charge the load capacitance, this is a fairly conservative value. We would also have to allow for the additional rise time as part of the timing analysis for the low-to-high transition.

Worst-Case Timing Analysis Example

Let's suppose an LSTTL gate is used to enable the D input of a flip-flop frequency divider, as shown in Figure 3-16. Figure 3-17 shows a functional timing diagram for the circuit in Figure 3-16, and Figure 3-18 illustrates a specification timing diagram for the same circuit. The timing of the input signals must conform to the combined specs of both devices, as defined below:

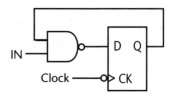

Figure 3-16: Example of worst-case timing.

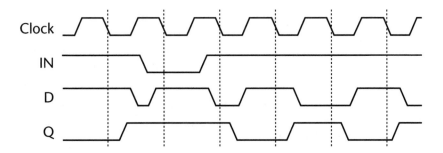

Figure 3-17: Functional timing diagram for Figure 3-16.

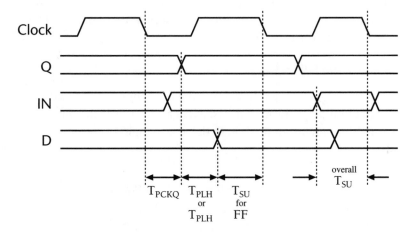

Figure 3-18: Specification timing diagram for Figure 3-16.

Flip-Flop Timing Specs

Symbol	min	typ	max	units
T_{SU}	10			nS
T_H	1			nS
T_{PCKQ}			15	nS
T_{PWCK}	10			nS
F_{CLK}			50	MHz

Gate Timing Specs

Symbol	min	typ	max	units
T_{PHL}	1	2	5	nS
T_{PLH}	2	4	6	nS

Test conditions $R_L = 1K$, $C_L = 100\ pF$

For the circuit shown in Figure 3-16 and the accompanying specifications, what is the maximum guaranteed clock rate?

From the timing figures on the previous page, note the minimum clock cycle time is defined by the sum of the following times: the time it takes for the transition from the active edge of the clock for the signal at D to propagate through the flip-flop, through the NAND gate and the time the signal must be stable before the next clock. The maximum propagation times and minimum setup times are used as they are the most severe requirements.

$$T_{PCKQ} + T_{PLH} + T_{SU} = 15 + 6 + 10 = 31 \text{ nS}$$

$$f = 1/t = 1/31nS = 32.26 \text{ MHz}$$

Now let's determine the setup and hold time requirements for the overall circuit. The overall setup time is lengthened by the delay of the NAND gate, therefore the system setup time is the sum of the flip flop setup time and the worst case propagation delay.

$$T_{SU}(\text{system}) = T_{PLH} + T_{SU}(\text{flip-flop}) = 16 \text{ nS minimum}$$

For the overall system hold time, the hold time of the flip-flop is offset by the minimum delay through the NAND gate, as this is the minimum amount of time that can be counted on to delay a changing D input to the flip-flop.

$$T_{H}(\text{system}) = T_{H}(\text{flip-flop}) - T_{PHL}(\text{min}) = 1 - 1 = 0 \text{ nS}$$

The delay in the D signal path reduced the hold time requirement from 1 nS to 0 nS, meaning the input can change at the same time as the clock edge or later. This is actually an improvement on the performance of the flip-flop by itself, which requires that the D line be held stable for 1 nS after the clock edge.

Chapter Three Review Problems

For the following problems, refer to the loading example and Figure 3-15.

1. If a 10 kilohm pull-up resistor is used, how many additional LSTTL loads can be connected?

2. How many CMOS loads could be added?

3. What could be done to increase the number of LSTTL loads?

For the following problems, refer to the timing example and Figure 3-16.

1. Using the same D flip-flop specified in the example, how fast could it be clocked if the /Q output was directly connected to the D input? (That is, eliminating the gate from the circuit.)

2. Under what conditions would the addition of a pull-up or pull-down resistor increase the fan-out of a logic output?

3. What, if anything, can be done to increase fan-out when it is limited by AC (capacitive) loading?

4. A 32-bit CMOS 5 volt microprocessor that has a 32-bit address bus and a separate 32-bit data bus, and the processor has a 1 nS rise time and 0.5 nH of ground inductance on a board made from glass epoxy material. The processor has output high and low voltages of 4.5 and 0.5 volts respectively and drives a capacitance of 100 pF on the address and data buses. How long can the printed circuit traces be before they must be considered as transmission lines?

5. For the same processor and conditions described in the last problem, what is the worst-case ground bounce voltage that can be expected?

Memory Technologies and Interfacing

Memory is one of the technology drivers in the integrated circuit business because the highly repetitive nature of memory arrays. Relatively small improvements in the design of a memory bit multiplied by the large number of bits on a chip can make a big difference in chip cost and performance. Gordon Moore, one of the founders of Intel Corporation, stated memory size doubles approximately every two years. The generalized version of *Moore's Law* (named after Gordon Moore, a co-founder of Intel who first articulated it) states that chip complexity doubles approximately every two years. As can be seen from Figure 4-1, as the resolution of features is reduced by a factor of 1/n, the area required for a gate is reduced by $1/n^2$. This exponential growth in complexity has continued in spite of those who have pointed out many reasons why it cannot continue. The supposed barriers have been overcome so far by various means to compensate for the limits of basic physics, such as pre-distorting the master patterns to compensate for optical diffraction effects.

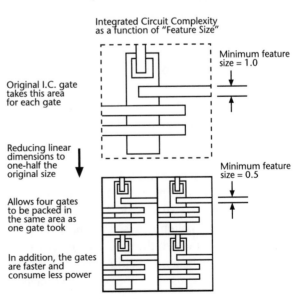

Integrated Circuit Complexity as a function of "Feature Size"

Original I.C. gate takes this area for each gate

Minimum feature size = 1.0

Reducing linear dimensions to one-half the original size

Allows four gates to be packed in the same area as one gate took

In addition, the gates are faster and consume less power

Minimum feature size = 0.5

Figure 4-1: IC density versus feature size.

The same technologies that were developed for memories have been applied to programmable logic and microcontroller chips. Each memory technology has unique advantages and limitations that the designer must be aware of. The wide variety of storage concepts and technology are central to the design of microcontrollers, and are categorized and described in this chapter.

Memory Taxonomy

There are many classes of memory devices, and the emphasis is placed here on those that are of significance to the designer of embedded systems. As a result, most of this chapter is dedicated to solid-state semiconductor memory chips rather than magnetic and optical storage devices.

The most significant distinction between memory devices is how they are connected to the CPU. There are two ways of connecting memory to the CPU:

- **Primary memory** the CPU is directly connected to the memory

- **Secondary memory:** connected to the CPU indirectly

Figure 4-2 illustrates the difference in the way the two types are connected to the processor bus. The CPU is only able to directly access information stored in primary memory. All instructions and data must be transferred to primary memory first before the CPU can process them. An example of primary memory is semiconductor RAM. The term RAM is frequently, but improperly, used to refer to primary storage. RAM only specifies the access mechanism (described below) but is often misused to imply the primary read/write semiconductor storage from which the CPU fetches instructions and data. Random access methods may be used in either primary or secondary memories, but are most commonly used for the primary storage, which is why RAM has been associated

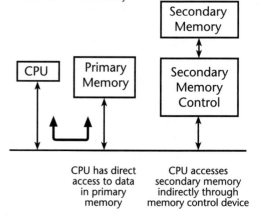

Figure 4-2: Primary versus secondary memory.

with primary memories. Because the CPU must access instructions and data quickly, primary memory must have very fast access time, on the order of tens to hundreds of nanoseconds or approximately 10^{-8} to 10^{-7} seconds, compared to secondary (disk) memory with memory access on the order of milliseconds (10^{-3} seconds).

Unfortunately, semiconductor memory, which is used for primary storage because of its high speed, is much higher in cost, size, and power per bit of storage than secondary memories. Semiconductor memory is currently the most practical mechanism for storing programs and data that are available for immediate use by the CPU. This is because the primary program and data memory must operate on the order of the speed of the processor memory cycles. Otherwise, the memory speed limits the overall system speed, because the CPU would have to be forced to wait until the memory is ready. One or more CPU clock cycles would have to be added to each memory access in order to slow the CPU down to match the speed of the memory. These delay cycles are referred to as *wait state*s because the processor must wait for one or more clocks before the memory data is available to the CPU.

Secondary Memory

A separate intermediate device usually controls secondary memory, which is not directly accessible to the CPU. The device manages the transfer of information between the storage device and the processor bus. When the data stored on a secondary memory device is needed by the CPU, it must first be moved to primary memory via the controller before the CPU can access it. Examples of secondary storage include magnetic and optical disk and tape that are used for large information stores because of their low cost per bit combined with high density and low power. Because of these differences, typical microcomputer architectures have about an order of magnitude larger secondary memories than primary memories. Secondary memories such as disk drives are most appropriate for storing large programs and data sets that must be maintained over a period of time. Secondary memories like magnetic tapes are often used for archival or backup storage because of their very high density and low cost. Another major advantage to magnetic and optical storage is that it is non-volatile.

Volatility

Non-volatile memories, such as magnetic disk and tape, maintain the information stored in them even when the power is removed. Volatile memories, however, do lose the information they hold when power is removed from them. The primary storage read/write RAM in a PC is volatile, which is why it must be reloaded with the operating system software (referred to as bootstrapping and loading the operating system) when the power is restored. In embedded controller designs, non-volatile memory is used to store the programs and constant data, and volatile memory is used to store the variables and temporary data.

Random Access Memory

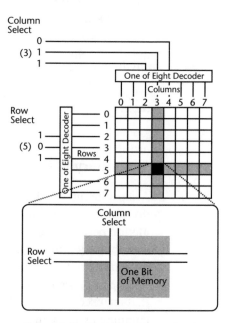

RAM is unique because the access time is essentially independent of where the data is stored. The random access method can be likened to the rows and columns of a spreadsheet, or the "pigeon hole" style boxes in an old desk. The specific memory location of interest is selected by a unique row and column address as shown in Figure 4-3. The row and column access can be used to select bits on a memory chip as well as chips on a memory board. Random access memory sizes are specified as 2^n x m, where 2^n refers to the number of

Figure 4-3: Random access memory (RAM).

unique locations or addresses and m is the number of bits stored in each location. A typical memory with 15 address lines and 8 data lines would be specified as a "32K x 8" or 32 kilobytes, since 2^{15} is 32,768 or 32 kilobytes. Another memory might be described as "4M x 1," meaning four million locations each containing one bit. Eight 4M x 1 memories can be wired in parallel to provide four megabytes of data for an 8-bit processor, or 16 can be paralleled to provide eight megabytes of data organized as 4M x 16.

Sequential Access Memory

Sequential access memory has an access time that is dependent upon the location of the data that is to be accessed. This is best illustrated by using the most common sequential access device: a magnetic tape. The information is stored in a serial fashion onto the tape, and the only data that can be accessed at any instant is the data stored on the tape in contact with the read/write head. Thus when the head is positioned at the beginning of the tape, the entire length of the tape must pass by the head before the last item can be accessed.

Direct Access Memory

Direct access memory which is a sort of combination of random and sequential access methods, is used on disk drives to provide an intermediate access time to fill the gap between high-speed random and low speed sequential access devices.

The storage medium is disk shaped, and contains a magnetic film for standard "hard drive" or fixed magnetic disks. Optical disks use an ultra-thin optical metal film that can be written once with a high intensity laser or read back using a low power laser. Optical disks that can be erased and re-written use a magneto-optical film whose optical properties (light polarization angle) can be changed using a low power laser and a magnetic field.

In each case, information is stored on concentric rings, called tracks on the disk. The information is stored sequentially on each track as it is on tape, but the read/write head can be moved to select the appropriate track. Disks with multiple recording surfaces also have multiple heads to read each surface, so they are randomly accessible by head and track, and sectors are sequentially accessed on each track. Figure 4-4 illustrates this.

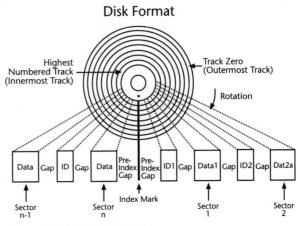

Figure 4-4: Direct access memory.

Another way of classifying memory devices is based on how information is written into the memory. *Read/write memories* are memories that can be written to as easily as they are read from by the processor.

Read/Write Memories

Static RAM or SRAM refers to a volatile semiconductor read/write memory in which the basic storage element is a flip-flop to store each bit. The flip-flops are arranged in rows and columns and are available in several organizations. The flip-flops take about four transistors per bit of storage, so they are generally about four times less dense than DRAMs that use only one transistor per bit. While these devices are volatile, they will maintain information as long as they are powered, unlike dynamic RAM that must be refreshed.

Dynamic RAM or DRAM, is a memory using a capacitor as the storage element. The presence or absence of charge on the capacitor represents ones and zeros. Because the capacitors are not perfect, they leak charge and will "forget" in as little as a few milliseconds if they are left alone, rather like a small child after being told to clean her room. In order to make the capacitors useful for storage they must be periodically *refreshed*. This is done by sensing whether there is any charge present on the capacitor and recharging the capacitor if there was charge present when it was sensed.

Refer to Figure 4-5. Charge is stored on the parasitic gate capacitance of a MOSFET transistor so that only one transistor is required per bit of storage. The process of reading or sensing the data is destructive in the sense that the charge representing the data is lost when it is sensed. The DRAM capacitor must be refreshed whenever it is read, and also periodically to restore the charge that leaks away. Each row in a DRAM has a sense amplifier and recharge circuitry designed to read and restore the data on an entire row at once. In order to refresh the DRAM data, a special abbreviated read cycle must be performed for each row of the memory. Because of the high density of data storage in DRAMs such as a 4 megabit device, the memory must have 22 address bits to select the location to be read or written. Rather than using 22 individual pins to specify the location, 11 wires are used and the address is latched by the DRAM in two parts: the row address and the column address. This is referred to as a *multiplexed address* Two control signals, row address strobe (RAS) and column address strobe (CAS, are used to multiplex the two 11-bit halves of the address into the DRAM. To simplify the refresh process, only the row address is used in a refresh cycle. Doing this takes advantage of the fact that there is one sense and refresh circuit for each bit in a row. The refresh row address is sequenced through all possible addresses before the capacitors can discharge.

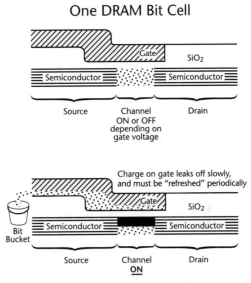

Figure 4-5: Dynamic RAM bit storage mechanism.

Read-Only Memory

Read-only memory (ROM) is a class of storage that cannot be erased or modified by the processor. Typical embedded systems may make use of one or more of the following types of ROM: mask ROM PROM, EPROM EEPROM or flash EPROM.

Mask ROM is memory that has been programmed at the time it is manufactured and can never be changed. The data patterns are defined by the photographic masks used to define the circuits on a chip when it is being fabricated. Mask ROMs are used when the programs or data do not need to be changed, when the production quantities are large, and the cost must be as low as possible. This is the oldest form of ROM and is still used in high volume applications because of its very low manufacturing cost. The program must be permanently defined in advance by including it as part of the master artwork film or "masks" used to fabricate the chips. It is also the least flexible to change, as a program change necessitates building and packaging new chips, which can take from weeks to months to accomplish.

PROM is user-programmable ROM, which is often used as a generic term for memories that can be programmed one or more times by the user using a special device called a *PROM programmer* or *PROM burner*. This was the first "field programmable" memory, meaning that it can be loaded with data by the end user using special programming equipment. Bipolar *fuse-link PROMs* were the first in this category, and were programmed by literally burning out fuses selectively from an array. This is where the term "burning" a PROM came from. (Up to now, you probably thought "burning a PROM" was some reference to the Stephen King novel "Carrie," didn't you?) Obviously one time programmable memory like this was expensive, since it was necessary to discard an obsolete device, and reprogram a new one every time a software revision needs to be tested.

Erasable PROM, or EPROM is used most frequently to store permanent data and programs. It is electrically programmable using an EPROM programmer, and can also be erased by shining a short wavelength ultra violet light through the transparent window in the IC package. The entire memory device is erased since it is not possible to be selective about where the light shines on the chip. These devices are also referred to as UV EPROMs. A *one-time programmable* (OTP) EPROM is simply an EPROM enclosed in a low cost package without a transparent lid, meaning it cannot be erased once it is programmed. The storage element in an EPROM is similar to that of a DRAM, as shown in Figure 4-6. However, the EPROM storage transistor gate is a conductor floating in an insulating SiO_2 (quartz) insulator, which prevents the charge from leaking off. The fact that the charge is generally guaranteed to remain for at least ten years in the absence of power—as long as the window is covered— makes this a non-volatile memory. This would be an ideal storage mechanism except for the way that the charge is stored on the gate. The charge is placed on the floating gate by a method called *avalanche induced migration*. This programming method is analogous to routing a river through the room to fill your cup with water. A relatively high voltage, 12 to 25 volts typically, is used to induce avalanche

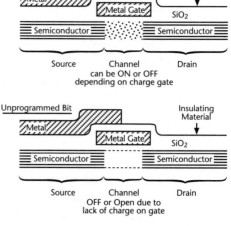

One EPROM Bit Cell

Figure 4-6: EPROM storage mechanism.

current flow across the insulating region for up to 50 milliseconds, and some of the charge is stranded on the floating gate. Figure 4-7 illustrates the program and read operations of a typical EPROM.

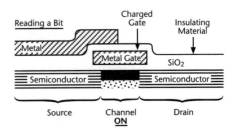

Figure 4-7: EPROM program and read operation.

EPROM erasure is accomplished by shining high-energy photons (UV light) onto the floating gates for several minutes, as shown in Figure 4-8. The photons impart enough energy to the trapped electrons to allow them to

escape the gate. The EPROM can be erased and reused many times, which is important when programs are in development, and when a reusable non-volatile memory is required. Some of the larger (less than 1 megabyte) EPROMs are available with a bank switching system to allow access to more locations than can be directly accessed using the address lines. This is accomplished using a write cycle to load the upper address bits into a latch inside the EPROM.

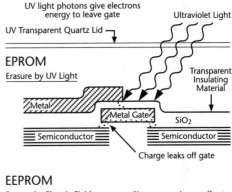

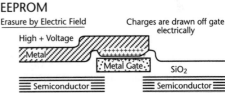

Figure 4-8: EPROM/EEPROM erasure.

Flash EPROMs are a variation on the standard EPROM, except that they have been modified so that they do not need to be exposed to UV light to be erased. Like an EPROM, the entire chip is erased at one time, but the erasure is performed electrically using a high reverse polarity voltage to remove the electrons from the gate. They are also easier to program and erase in the application design using relatively simple additional support circuits.

EEPROMs, or *E²PROMs*, are *electrically erasable* PROMs. They can be erased and written electrically one byte at a time. The mechanism used is similar to the EPROM except that the insulating region is made very thin, on the order of a few angstroms. The charge is transported using an effect referred to as

Fowler-Nordheim tunneling where the insulator is thinned. In an interesting application of quantum physics, the electrons "tunnel" through the insulator. The operation is similar to an EPROM except that most types can be erased and programmed in circuit, using 5 volt power supplies and a standard microprocessor bus interface. For many of the devices, each byte must be erased by writing ones to a location before it can be programmed. While these devices would seem to be nearly ideal as non-volatile read/write memories, they do have a couple of drawbacks. EEPROM bits have a limited number of write cycles before they get "stuck" in the programmed state. They are typically guaranteed for 10,000 to 100,000 write cycles, which would take only a few seconds if a program gets stuck in a tight loop writing to the EEPROM. This problem is due to the fact that charge can be trapped in defects in the insulator in the gate region resulting in some bits getting "stuck" in the programmed state. The other problem is that they are slow to write, typically taking many microseconds or even milliseconds to erase and write, compared to 100 nanoseconds typical of SRAM.

Small EEPROMs are available with a serial interface so that they will fit into small (8-pin), low cost packages. They are particularly useful in embedded systems for storing configuration data to replace switches and jumpers. They are significantly slower than standard memories due to the serial interface, and are usually accessed using software to manipulate the serial lines directly.

Other Memory Types

Battery-backed CMOS SRAM or NVRAM (*non-volatile RAM*), is a device consisting of a low power CMOS SRAM, a battery, and control circuitry to maintain the data in the RAM using the battery when the external power is off. These devices come in two forms: an oversize IC socket containing the battery and control circuit into which a CMOS RAM is inserted, and a single package containing all three components with the RAM permanently installed. The advantages of these devices are that they have the same ease of read and write as a standard RAM, unlimited read/write cycles, and non-volatility. Disadvantages include the environmental and storage life limitations imposed by the battery, and delayed access to the data during power application. Write cycle access is disabled for a fixed period of time after the power supply reaches a predetermined voltage to prevent spurious write signals from corrupting the data while the processor is in the process of initializing itself. As a result of this initial period when the memory is "write-protected," the processor cannot store data such as subroutine and interrupt return addresses on the stack. This can result in unpredictable operation unless the software has been designed to allow for the necessary delay after power up to guarantee that the memory will accept a write cycle.

Ferro-electric RAM is a semiconductor memory with a combination of fast access, unlimited write cycles, and non-volatility. This is a relatively new and unproven type of device that stores information using a material that can change its properties and be sensed electrically, but retains its data like magnetic storage when power is removed. The cost of these devices is high relative to the other types, however, limiting its potential applications.

The proliferation of memory technologies is due to the compromises in current memory devices. The ideal memory would be low cost, high density, random access, fast access time, read/write, and non-volatile, with unlimited read/write cycles. Each of the memory devices discussed is optimized to incorporate several of these ideal characteristics at the expense of the others. Because of these compromises the designer most often uses multiple types of devices to meet conflicting memory requirements. Probably the most common solution for embedded processors is the use of EPROM to store programs and constant data, and SRAM to store read/write data such as variables and stacks. EEPROMs are becoming more popular in embedded designs because they allow storage of information that is infrequently updated such as calibration and configuration information.

JEDEC Memory Pin-Outs

RAM, ROM, EPROM, and EEPROM pin-outs have been standardized to make it easier to design a microcomputer memory interface. The JEDEC (Joint Electronic Device Engineering Committee) standard defines the pin-out of the devices so that various types of memories can be installed at the same site in a circuit board with a few jumpers. This standard encompasses 24-, 28-, and 32-pin DIP devices, as well as equivalent surface mount packages. As a result, the data lines, and many of the address and control lines, are unchanged for a wide range of device sizes and memory types.

Figure 4-9 shows the pin assignments for a 32 kilobyte EPROM in a 28-pin DIP package and a 128 kilobyte static RAM in a 32-pin DIP package. Note the commonality of the two assignments. Both types of devices can be accommodated in the same pattern on a circuit board by connecting the common pins directly to the appropriate signals, and by providing movable jumpers or programmable logic to allow use of either type of memory. This particular pin-out pattern, or "footprint," is standardized by JEDEC and is referred to as the JEDEC 28- or 32-pin *memory footprint.*

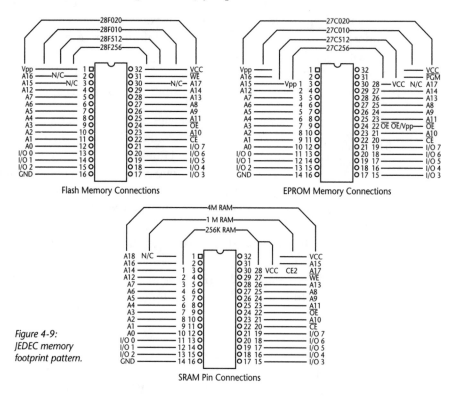

Figure 4-9:
JEDEC memory
footprint pattern.

Device Programmers

A special device is required in order to program most types of PROMs because of their signal timing and voltage requirements. The *PROM programmer* or *PROM burner* programs a device with the data pattern from a master PROM, a serial port, or a disk file. "PROM burner or blower" is a term that originated when programming fuse-link PROMs required that the fuse be burned or blown to program each bit in the memory. Device programmers come in two forms: a desktop instrument with serial I/O ports and disk drive for data input, or a PC compatible plug-in card. The desktop units are more versatile, but the plug-in cards are much less expensive. The most flexible units in both categories have a programmable power source and sense circuit on each device pin, and the least expensive are those that program only one type of device. Some programmers are also capable of programming PLDs (*programmable logic devices*) in addition to standard PROMs.

The procedure for programming an EPROM that has been used before is typically:

1) Remove the label covering the quartz window on the EPROM.

2) Place the EPROM in a UV EPROM eraser for 20 minutes to erase existing data.

3) Turn on the EPROM programmer.

4) Select the type of device to be programmed.

5) Load the data pattern into the programmer from a computer using a serial port.

6) Put the EPROM into the appropriate socket in the programmer.

7) Start the programmer, and wait anywhere from few seconds to twenty minutes.

8) The programmer indicates that the EPROM is properly programmed.

9) Remove the EPROM and cover the window with an identifying label.

10) Install the EPROM in the circuit board where it will operate.

Procedures for each programmer vary in detail, but the overall process remains the same. The photo below shows what one type of programmer looks like. A *zero insertion force* (ZIF) integrated circuit socket is used to make it easy to insert and remove the device and to prevent damage to the pins on the device. Programmable devices in surface mount packages will also require the use of a special adapter. Programmers for memory and programmable logic devices are available at prices ranging from hundreds to thousands of dollars.

Memory Organization Considerations

A generic memory device is organized as a number (N) of locations multiplied by megabits per location. Here are two examples:

- "128K x 8" SRAM representing a chip with 128,000 locations of eight bits each = 128 kilobytes of storage capacity.

- "1M x 1" DRAM with 1 million locations each with one bit = 1 megabit of storage.

Note that *both of these memories have one million bits of storage*; they are just organized differently. Figure 4-10 illustrates this. The SRAM has eight bits per location and is referred to as a *byte-wide* memory, and the DRAM has one bit per location. The SRAM chip has $\log_2(128 \text{ kilobytes}) = 17$ address bits numbered A0-16 and eight data bits numbered D0-7, where A0 and D0 are the LSBs (*least significant bits*). Because the address bits are multiplexed to share address pins, the DRAM has $\log_2(1M) = 20$ address bits or $\log_2(1M)/2 = 10$ address pins.

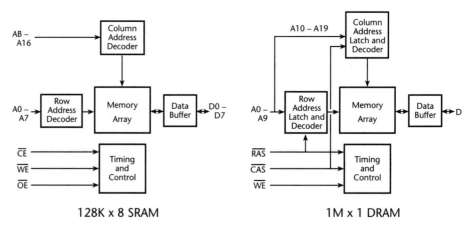

Figure 4-10: Two different memory organizations for 1 megabit memory.

Byte-wide memories are more common in microcontroller designs due to their simplicity and the wide variety of memory technologies available in standard JEDEC pin-out packages. While byte-wide SRAM memories have a higher cost per bit-on-a-chip basis than DRAMs, they do not require any support circuitry for refresh and address multiplexing. For a system incorporating a small amount of SRAM, the overall cost and complexity are less than they would be for a comparable DRAM design. For RAM memories consisting of many chips, the DRAM's lower cost per bit outweighs the cost of the support circuitry. In order to increase density, DRAMs are often packaged in SIMM (*single in-line memory module*) form, which is essentially a very small circuit board containing eight or nine DRAMs which are then plugged into card edge sockets on the main board. This concept is very popular for PCs, workstations, and other general-purpose designs with requirements for large RAM storage.

Parametric Considerations

Timing parameters were discussed in detail in Chapter Three. However, there are several that are unique to memory devices. These include *access time*, *cycle time*, and, in the case of DRAM, *refresh interval*.

Figure 4-11 shows a timing diagram illustrating memory read cycle timing parameters. These access times include:

- T_{AA} (address access time): Valid Address to valid data delay
- T_{OE} (output enable access time): Output Enable (OE) to valid data delay
- T_{CE} (chip enable access time): Chip Enable (CE) to valid data delay

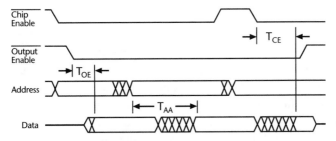

Figure 4-11: Memory read cycle timing parameters.

Figure 4-12 shows a timing diagram illustrating memory write cycle timing parameters. The pulse width, setup, and hold times include:

- $T_{WP:}$ Write pulse width
- $T_{AS:}$ Address setup time
- $T_{AH:}$ Address hold time
- T_{DS}: Data setup time
- T_{DH}: Data hold time

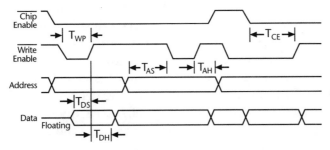

Figure 4-12: Memory write cycle timing parameters.

In addition to the memory timing specs shown above, some memories, such as DRAM, have additional constraints as follows:

Cycle times

T_{RC} (read cycle time): how closely read cycles can be spaced

T_{WC} (write cycle time): how closely write cycles can be spaced

Read-modify-write cycle time is a special combined read/write cycle to the same address (e.g. increment a memory location)

DRAM Refresh Cycle

T_{REF}: the maximum time between refresh/read/write cycles before DRAM data loss can occur

One of the DC characteristics of interest in an embedded system is the power consumption, particularly in a battery-operated design. Most static memories have low power or power-down modes activated by disabling the chip select or chip enable line. True CMOS SRAMs have typical power down supply currents in the low or sub-microampere range, allowing their data to be maintained using a battery while the main power is off as in an NVRAM. Some SRAMs are advertised as CMOS, even though they have some NMOS circuits internally to improve speed. These "mixed MOS" designs draw significantly more power and are not usually appropriate for typical battery operated applications.

Practical examples of actual memory specifications used in design of an embedded system can be found in Chapter Six.

Asynchronous vs. Synchronous Memory

An asynchronous memory is one that does not require any clock signals and delivers its output with a delay of one access time (the internal memory logic propagation time) after the address and control lines stabilize. Most SRAMs, like the SRAM described above, are asynchronous, but a few are synchronous and have clocks for internal latches to store the address and write enable signals. DRAMs are synchronous because they require RAS and CAS strobes to load the internal data latches. Generally asynchronous parts are easier to design with because of simpler timing constraints and direct compatibility with most processor buses.

Error Detection and Correction

Error detection circuitry stops an operation before erroneous data is used, such as a parity error trap. *Error correction* on the other hand, uses redundant data to reconstruct the original data to be used when operation must continue without interruption. Error detection and correction are not often used in small systems because of the relatively low probability of error and high cost of error detection and correction hardware. In systems like PCs and workstations, larger RAM memories result in the need for error detection as a minimum, and error correction in systems requiring high reliability. In most PCs, a ninth bit in each byte stores parity information, and if there is a parity error, an interrupt trap will stop operation and display an error message.

There are two types of errors: *hard errors* and *soft errors*. If an error occurs only once, due to noise or a transient error condition, it is referred to as a soft error. A hard error is one that always occurs, such as a read/write memory bit that is stuck in one state and can't be changed.

Error Sources

Hard errors are usually caused by a permanent hardware defect, while soft errors can be caused by any one of several events, including timing errors, synchronization problems, software bugs, or even the passage of a charged subatomic particle resulting from the decay of trace radioactive materials flying through an IC. As a designer of an embedded system, it is necessary to allow for the occurrence of these events, and minimize the severity of their effect on the overall system. In order to accomplish that goal, it is necessary to detect the occurrence of such an event as a minimum.

Confidence Checks

The *confidence check* is frequently used to detect these errors, and can be modified to correct certain subsets of the errors as well. Probably the most well known of the detection techniques is *parity*. Its widespread use is due to the simplicity of its implementation. In the most common form, a single bit is added to every word, containing the parity check bit. The parity bit is set or cleared depending on whether there are an even or odd number of ones in the original word to be checked. Whenever the data is handled, the contents are checked against the parity bit. If any one bit in the word has changed, then the parity of the data will not match the parity bit accompanying the data, indicating an error. For a single byte or word, this is usually a reasonable assumption, however for a large block of data, it is not reasonable. *Horizontal parity* refers to the parity of a single word of data, while *vertical parity* refers to the parity of one bit position in multiple words. They are combined to form *block parity*, which assigns one parity bit for each word horizontally and one parity bit for each bit position in the block of words.

Block parity allows the detection and correction of single bit errors. Since a single bit will cause one horizontal and one vertical parity error to occur, correcting the bit in error requires only complementing the bit belonging to the row and column corresponding to the parity errors. Note that multiple errors may not be corrected or even detected, depending on where they occur.

Here is an example using odd parity:

data:	Horizontal parity:
1 0 1 1	p=0 odd horizontal parity
1 1 1 1	p=1 even horizontal parity +1 = odd parity
1 0 0 1	p=1 even horizontal parity +1 = odd parity
1 0 1 1	p=0 odd horizontal parity
1 0 0 1	< The odd vertical parity bits for the four words above

Another version of parity checking is called *Hamming code* after its inventor, R.W. Hamming. It is a code in which multiple parity bits are appended to each word in such a way that a single bit error will generate a group of parity bits having a value equal to the data bit number in error.

A *checksum* is another technique that can be used to detect an error in a group of characters. The idea is simple enough: sum all the data words and keep the least significant bits of the sum. (For you math majors, that's summing the data modulo 2^n, for n bit words.) Checksums are frequently used by various types of memory and logic device programmers to verify that the desired program has been "burned" into the device. A checksum will detect some, but not all, of the common errors in a block of data. For example, it won't detect errors due to the data being stored in the wrong sequence, since the sum of the numbers is the same regardless of the order. A practical example is when a 16-bit CPU's program is burned into two 8-bit memories, one containing the lower byte and one containing upper byte of the instructions. When the bytes in the block of memory are summed, the answer is the same, even if the two devices are swapped! Thus, a serious and common error would not be discovered.

The CRC (*cyclic redundancy code*) is used to detect changes within a block of data or its order. The CRC is based on a polynomial that is calculated using shifts and XOR (exclusive OR) logic to generate a number that is dependent on the data and the order of the data. The detailed operation of a CRC is beyond the scope of this book, but is based on the same polynomials used for generating pseudo-random numbers. It is commonly used for checking blocks of data on magnetic storage devices and communication links.

Memory Management

In order to understand what memory management is, it's helpful to understand the motivation behind its use. There are two kinds of memory management: memory address relocation and memory performance enhancement. They are often used in conjunction, as is commonly done in personal computers. This section covers the performance enhancement aspects, while the address relocation issues will be covered in Chapter Six.

The differences between different storage technologies, in terms of performance and cost, vary over many orders of magnitude. For example, semiconductor memory devices have access times that are many orders of magnitude faster (nanosecond vs. millisecond access time) than that of magnetic disks. Of course, magnetic disks also have a cost several orders of magnitude less than semiconductor memory on a cost per bit basis. This disparity in price and performance has lead to the idea of using small, fast memories to store the most frequently accessed subset of the complete collection of data present in a larger, slower memory. This technique of buffering, often referred to as *caching* memory contents in a fast memory, is essentially similar whether it is applied to the memory attached to a CPU or the magnetic or optical storage mechanisms. In fact, there may be several layers of caching in a given system, starting with the smallest, fastest memory closest to the CPU, followed by slower but larger memories.

Memory price is inversely proportional to speed, as indicated below:

Memory type	Relative Access size(Bytes)	Relative Time(Sec)	cost/byte
Tape	10^{10}	10	1
Disk	10^9	10^{-3}	10
DRAM	10^6	10^{-7}	10^2
SRAM	10^5	10^{-8}	10^3

Cache Memory

When a high speed memory is used to provide rapid access to the CPU for most frequently used portion of main memory, it is referred to as a *CPU cache* memory. Likewise, when the main memory is used to provide rapid access to data stored on a disk, it is referred to as a *disk cache*.

The objective of these approaches is to maximize the likelihood that most pieces of data will be found in the small and fast memory most of the time, thus reducing the average effective access time. The object is to succeed at finding most data in the small fast memory most of the time, minimizing the number of accesses to the big slow memory. Fast SRAM is used as a fast temporary buffer (memory cache) between main memory and the CPU. Main memory DRAM is used to buffer disk data (disk cache). Most hard disk drives also have some internal fast semiconductor RAM to cache data as it is being transferred to and from the disk.

Virtual Memory

Disk storage can be used to emulate a larger primary memory than is actually available. *Demand paged virtual memory* provides an apparently large primary memory by swapping pages of data between real primary memory and disk. This is a combination of hardware for translating logical (virtual) addresses, moving pages as needed, and operating system software to determine where and when pages should be kept and detect access attempts to pages which are not in primary memory.

When address relocation mechanisms are combined with disk caching and special system software, it is possible to make the main memory appear much larger than it actually is to a program running on this type of machine. When the program attempts to access a location that is not present in the main memory, the hardware and software redirect the memory reference to a real block of memory, after the required data is loaded from disk. Thus the application program is presented with a virtual memory that is significantly larger than the actual physical main memory. This has the effect of simplifying the code, since all data can be referenced by a single address, rather than selecting a file, track, or sector on a disk.

CPU Control Lines for Memory Interfacing

Some CPUs generate signals for memory timing and synchronization with devices having various access times using a technique that generates delay cycles for slow memories, referred to as *wait states*. The 8051 processor used in this text does not use or generate wait states for simplicity. The Dallas 80C320 series of high speed microcontrollers incorporate a software-controlled mechanism for generating wait states. These extended memory cycles allow the processor to work with slower memory and peripheral chips.

Chapter Four Problems

1. What is the largest capacity SRAM that will fit in a 32-pin package?

2. What is the largest ROM that will fit in a 32-pin package?

3. Using 4M x 4 DRAMs, how many chips will be required to implement a 16 megabyte memory organized in 32-bit words?

4. What restrictions must be considered, when writing software to program an EEPROM device?

5. What restrictions are imposed when writing to flash EPROM?

6. What would you expect to read from a blank EPROM, if its data storage element is an N-channel FET that is connected with its source grounded and the drain connected to an output pin and a pull-up resistor?

CPU Bus Interface and Timing

The central processing unit (CPU) is the key part of a microcomputer, both from the functional aspect and from the design procedure facet. This is because the key control signals originate from the CPU, driving most of the timing, load, and functional characteristics of the bus interface that all other devices must be compatible with. The processor controls the data transfers on the bus on a cycle-by-cycle basis, fetching instructions, reading and writing operand data. Let's begin by examining how the CPU reads data from and writes data to memory.

Read and Write Operations

Refer to Figure 5-1 as you read through the following steps in a memory read operation:

1) The CPU selects the memory location by driving the address on the address bus.

2) Control lines are driven by the CPU to indicate the address space to use, such as program memory, data memory, I/O, or special cycles such as interrupts.

3) Read is activated on the control bus by the CPU to indicate that the memory can drive the data bus with the contents of the selected location.

4) The memory drives the contents of the selected location on the data bus.

5) The CPU deactivates the address and control lines, turning off the memory drivers.

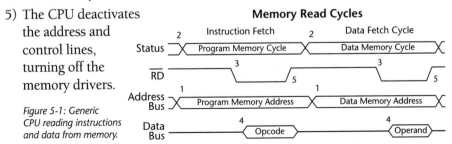

Figure 5-1: Generic CPU reading instructions and data from memory.

Refer to Figure 5-2 as you read through the following steps in a memory write operation:

1) The CPU selects the memory location by driving the address on the address bus.
2) Control lines are driven by the CPU to indicate the address space to use.
3) The CPU drives the data to be written on the data bus.
4) Write is activated on the control bus by the CPU to indicate that the data on the data bus should be written into the selected location.
5) The CPU deactivates the address, data, and control lines.

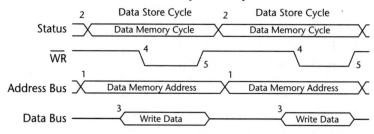

Figure 5-2: Generic CPU writing data to memory.

Address, Data, and Control Buses

During normal operation, the CPU drives the address bus with the location to be transferred to or from the CPU. Addresses generally refer to memory locations or I/O locations. The data stored in those locations is usually eight bits (a byte), 16 bits, or 32 bits depending on the processor. Most microcontrollers use byte addressing, meaning that each address is a pointer to an 8-bit piece of data. Most 8-bit and virtually all 16- and 32-bit processors can also address and manipulate data in 16- and 32-bit pieces. Directly accessible addresses are those that the CPU can access in a single cycle using the address bus. If a processor has N address bits, then it can *directly address* 2^N locations, starting at location 0 and increasing to location 2^N-1. Typical processors may have 16-, 20-, 24-, or 32-bit address buses. A byte addressing, 16-bit processor can address 2^{16} locations, or 65,536 = 64 kilobytes. Likewise, a processor with a 20-bit address bus can directly access 2^{20} locations, or one megabyte. Some locations of memory may not be directly accessible by the CPU, meaning that the CPU must use multiple cycles to access one memory location, usually under software control. This technique, sometimes referred to as *bank switching*, is the so-called "expanded memory above one megabyte in the PC, which uses an 8088 CPU with 20 address bits.

The 80286 CPU has 24 address bits allowing direct addressing of 2^{24} or 16 megabytes. The 80386 and higher processors have a 32-bit address space, addressing up to 2^{32} or 4 gigabytes. Some processors use a subset of the address lines for I/O. If the processor instructions use a 16-bit address field in the I/O instructions for example, then only 2^{16} I/O locations are accessible.

The data bus, driven by the CPU during write cycles and by other devices during read cycles, transfers instructions and data in and out of the CPU. The width of the data bus, among other things, determines the amount of data that can be transferred on the bus. This data throughput is referred to as the *bus bandwidth* and is usually expressed in bytes per second. If a bus supports one transfer per microsecond, an 8-bit bus has a one megabyte per second bandwidth, a 16-bit bus has a two megabytes per second bandwidth, and a 32-bit bus has four megabytes per second bandwidth. In the case of an 8-bit bus and a period T =1 microsecond (μS), then f = 1/T = 1 MHz and, for one byte per cycle, the result is one megabyte per second or eight megabits per second.

The control bus, normally driven by the CPU, determines what type of cycle is to take place and when the data will be present on the bus. In the case of a processor with a multiplexed address and data bus, some or all of the data bus is multiplexed or shared with the address bus. An additional signal is provided on the control bus to enable an address storage latch to hold the address information at the beginning of a transfer cycle. Bus cycles on a multiplexed address/data bus system, as shown in Figure 5-3, are identical to those illustrated previously except for the addition of address information on the data bus at the beginning of a cycle, and an address latch control signal as shown in Figure 5-3. The 8051 has a multiplexed bus cycle.

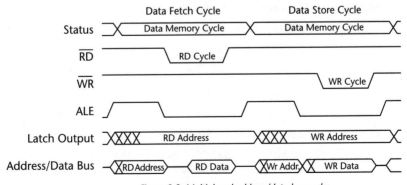

Figure 5-3: Multiplexed address/data bus cycles.

As soon as the address latch enable (ALE) is high, the address latch allows the multiplexed address from the address/data bus through to the latch output. When the ALE signal goes low, the address remains frozen on the latch output, and the CPU can remove the address lines from the bus and begin a data transfer.

The address latch must be a transparent latch with active high enable, such as the 74xx373 device. Figure 5-4 shows a typical arrangement. It is important to recognize that a transparent latch operates differently than a clocked register. As long as the '373 latch enable input is high, the latch Q output follows the D input. As soon as the latch enable goes inactive, the latch Q outputs freeze. This is analogous to the way a VCR allows a continuously changing signal show on the display until the pause button is pushed. This is in contrast with edge sensitive devices, such as the '374, which only updates the Q outputs at the *rising edge* of the clock. The '374 is analogous to a flash still camera, which captures the input at the instant that the flash occurs. If the ALE signal was inverted, the '374 latch would sample and hold the address at the end of the ALE pulse. While this could function correctly, it would delay the availability of the address to the memory devices, leaving less time for them to access the addressed location.

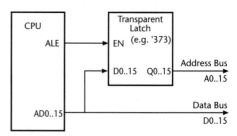

Figure 5-4: Address demultiplexing with a latch.

Address Spaces and Decoding

Processors, depending upon the particular architecture, may have several separate *address spaces*, such as the following:

- program memory address space
- data memory address space
- input/output device address space
- stack address space

Depending on the processor, these may be completely separate, overlapping, or all-in-one address space. When these are separate spaces, the processor has separate control signals to indicate which address space is to be used for data transfer. This may be done with a separate signal line that goes active when a particular space is being addressed, such as a program fetch denoting that the data should be transferred from a program memory address. The address space selection may also be performed using several status lines that, when decoded, define the appropriate transfer as in the case of the Intel 80x86 family. When there are separate address spaces, as in Harvard architecture CPUs like the 8051 family, there will be more than one unique location with the same address. The status and control lines are needed to single out the appropriate location as shown in Figure 5-5.

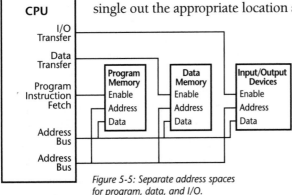

Figure 5-5: Separate address spaces for program, data, and I/O.

Some processors, such as those in the Motorola 680x0 family, have a single address space for all purposes, including I/O. Dedicating part of the memory address space to I/O is referred to as *memory mapped I/O*. Even processors that have separate I/O instructions and address space may have some memory mapped I/O by dedicating some of the memory address space to I/O devices.

The various address lines and control lines are decoded to provide individual chip select signals for the various memories and I/O chips. This is the purpose of the *address decoder*. A standard n-line to 2^n-line decoder is sometimes used to decode the address lines. A typical device is the 74LS138, a 3-to-8 line decoder that drives one of eight output lines low, depending on the three bit binary number on the input. For example, with 16 address lines there are 64K unique locations in a memory address space. This would require eight memory ICs if each one contains 8K locations (64K locations divided by 8K locations per chip = 8 chips). By connecting the three decoder inputs to the most significant bits of

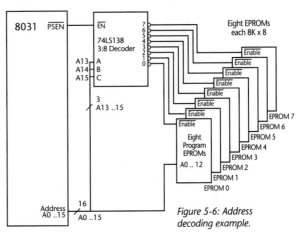

Figure 5-6: Address decoding example.

the address bus and each of the eight decoder outputs to a memory IC chip enable, one of the eight memory devices will be selected for any given address. Decoders also have enable inputs that can be used to enable the outputs only for a selected address space such as memory or I/O. The example in Figure 5-6 shows an 8031 with eight program EPROMs.

Address Map

In order to describe the address decoding of memory and I/O clearly an *address map* (also referred to as a *memory map*) table is used to specify which devices respond to a particular range of addresses in a given address space. The purpose of an address map is to clearly define the range of addresses that each memory or I/O device occupies in the address space. A separate map is used for each address space in processors that have more than one address space. For example, the 8031 has a factory defined map of the internal data memory address space, another map for program memory, and a third for external data memory. It also helps to define which memory space any given device resides in. As an example, the address decoding table for Figure 5-6 is shown in Table 5-1:

Address Range (hex)	Address bits A15 A14 A13	Decoder Ouputs 76543210	Chip Select Active for Memor y IC
0000 - 1FFF	0 0 0	11111110	EPROM 0
2000 - 3FFF	0 0 1	11111101	EPROM 1
4000 - 5FFF	0 1 0	11111011	EPROM 2
6000 - 7FFF	0 1 1	11110111	EPROM 3
8000 - 9FFF	1 0 0	11101111	EPROM 4
A000 - BFFF	1 0 1	11011111	EPROM 5
C000 - DFFF	1 1 0	10111111	EPROM 6
E000 - FFFF	1 1 1	01111111	EPROM 7

Table 5-1: Memory map for Figure 5-6.

The same decoding technique can be applied to I/O devices to select one of several devices. In the case of an I/O decoder connected to a processor with a separate I/O address space, the decoder's enable input would be controlled by the CPU I/O control line. Whenever an I/O cycle occurs, the I/O device address is presented on the address bus and the I/O control line is activated. This causes one of the decoder outputs to go active and select an input or output port. In the case of memory mapped I/O, the decoder outputs would go to both memory and I/O devices. An I/O address map is used to specify the location(s) in I/O address space that each device will respond to. The map may also specify if the location is read only, write only, or read/write. This is because I/O device addresses are not always read and write. As an example, an output port that drives some LEDs would be an output only or "write only" port. Microcontroller chips usually have some dedicated input bits and output bits as well as some general purpose I/O port bits implemented directly on the chip, which are usually accessible by reading or writing special register addresses. Microprocessors and microcontrollers with external buses can also have memory mapped I/O. The example below shows a one bit input port and a 1-bit output port mapped into the external RAM space along with six 8Kx8 RAMs.

The example address map in Table 5-1 and decoder circuit in Figure 5-6 illustrate complete address decoding. That is, there is one device mapped to each block of addresses in such a way that all the addresses map to one and only one unique set of memory locations. Each of the eight memories containing eight kilobytes of memory maps to one of the eight regions of eight kilobytes. There are no unused addresses, and there are no duplications. If all possible addresses are decoded, but some are not used, then it is possible to expand the memory available by using the available memory address ranges for additional memory. If any device is decoded in such a way that it appears more than once in the address space, then it is referred to as *partial address decoding*. This derives from the fact that not all the address signals are used to determine which device should be enabled. This is often done to reduce the complexity of the decoding circuits, at the expense of future expansion options.

In the address decoder shown in Figure 5-7, the I/O addresses are partially decoded, resulting in a range of addresses that enable a single device (the I/O port).

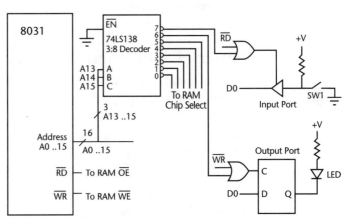

Figure 5-7: Memory mapped I/O in the 8031 external memory space.

Note also that two separate address ranges have been used, one for the input port and one for the output port. In practice, it is possible to have the input and output ports respond to the same address by using the read line for input cycles, and the write line for outputs.

Address Range (hex)	Address bits A15 A14 A13	Decoder Ouputs 76543210	Active Select: Memory I/O
0000 - 1FFF	0 0 0	11111110	RAM 0
2000 - 3FFF	0 0 1	11111101	RAM 1
4000 - 5FFF	0 1 0	11111011	RAM 2
6000 - 7FFF	0 1 1	11110111	RAM 3
8000 - 9FFF	1 0 0	11101111	RAM 4
A000 - BFFF	1 0 1	11011111	RAM 5
C000 - DFFF	1 1 0	10111111	Output Port
E000 - FFFF	1 1 1	01111111	Input Port

Table 5-2: External data memory map (8031 external memory space).

The decoder will select the input port at *any* address in the range E000 through FFFF hex. That means that the single input port bit takes up 8K address locations, all reading the same input port. This decoding technique is partial address decoding because only the three most significant address bits are decoded for this input port, and the rest of the address lines are effectively "don't cares." This may seem wasteful of address space, but it reduces the amount of decoding circuitry when it is not necessary to decode all the unique addresses individually. The memory map of the external data memory address space is shown in Table 5-2.

Chapter Five Problems

1. If the design of Figure 5-7 needs to be changed to eliminate the duplication of addresses caused by partial address decoding, how many additional input signals would be required for the decoder?

2. The 8031 CPU has 16 address lines. How much external memory can be attached to it without resorting to any memory extension mechanism?

3. If all bits of Port 1 on an 8031 are used to select external data memory in one of 256 "banks," what is the maximum amount of external data memory that can be accessed?

4. What is the answer to life, the universe, and everything?

A Detailed Design Example

In this chapter, we will take a detailed look at the design and analysis of a simple microcontroller project. This chapter will illustrate the interactive nature of the design process. First, the preliminary design is analyzed for limitations and violations of the timing requirements for the various chips. Then modifications and additions to the design are made to improve the performance based on the analysis. The modified design is then verified for conformance to the various component specifications. This iterative process begins with a simple block diagram showing the components of interest and progresses to detailed timing diagrams, specifications, and timing analysis.

The Central Processing Unit (CPU)

The process of designing an embedded microcomputer system is mostly independent of the particular CPU that is used. The example design of this chapter is a relatively simple one that illustrates the design and analysis process in enough detail to show what needs to be done. Because the Intel 8031 micro-controller design has a simple bus interface, has brief timing specifications, uses SRAM, and incorporates relatively simple I/O on chip, it will be used to illustrate the critical design and analysis processes. Once the complete process is understood with this simple CPU, more advanced designs can be addressed with comparative ease.

The 8031 processor is a Harvard architecture with a multiplexed address and data bus. There are three address spaces: internal RAM, external data RAM, and external program ROM. The external program ROM and data RAM are accessed using three memory cycles: program read, data read, and data write. Three separate, mutually exclusive control signals from the CPU determine which of the three types of external memory cycle are to occur. Only one of the signals is active at any one time, making the memory interface very simple. A program read cycle is indicated when the /PSEN (active low, program strobe enable becomes active, a RAM data read cycle when /RD (active low, read) goes active, and a RAM data write cycle is indicated when /WR (active low, write) becomes active. The /PSEN signal can be directly connected to enable the program ROM, and the /RD and /WR signals can be connected to the output enable and write enable pins of the data RAM. Since the lower eight address bits are multiplexed on the data bus, they are held by a transparent latch (74x373). The processor outputs an active high enable signal, ALE (address latch enable), to control the latch. The processor, latch, program EPROM, and SRAM are shown in Figure 6-1. The timing diagrams

for the three memory cycles as shown in the processor specification, along with the timing parameters for the CPU, are shown in Figure 6-2. The CPU timing requirements must be reconciled with the requirements of the other chips in the system, beginning with the memory chips.

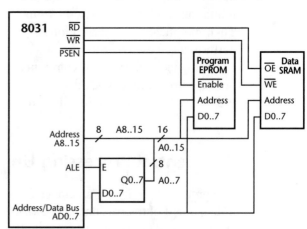

Figure 6-1: Preliminary design of the CPU and memory interface.

Memory Selection and Interfacing

Most embedded computer designs make use of EPROM for non-volatile program storage and SRAM for volatile data storage. For this example we will use one of each type: 32Kx8 UV erasable EPROM to store the program, and a 32Kx8 CMOS static RAM. The multiplexed address bits, A0..7, will be latched from the AD0..7 lines using a 74ALS373 transparent latch. Since there is only one memory of each type, no address decoding is necessary for the chips to be enabled directly from the processor memory control lines /PSEN, /RD, and /WR.

Preliminary Timing Analysis

Critical timing parameters for the EPROM, SRAM, and address latch are shown in Tables 6-1 and 6-2 and are excerpted from the component specification sheets. For an experienced designer, the preliminary timing analysis may consist of just a quick look at the data sheets. A limited analysis of key timing parameters will be performed first to identify any major changes that may need to be made in the design. The parameters to be evaluated here will be the CPU memory access time requirements versus the various memory maximum access time capabilities, control signal pulse widths, all related to clock speed. First, we will examine the program memory read access time, and then the data read and write access times.

The instruction fetch (program memory read) cycle of the CPU is shown in Figure 6-2.

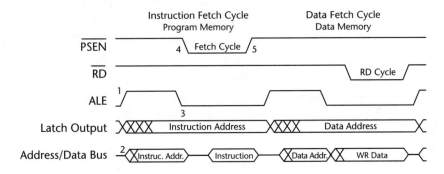

Figure 6-2: Instruction cycle timing diagram.

The sequence of events is as follows:

1) ALE goes active (high), enabling the external latch to pass A0..7 through to its outputs. The 16-bit PC (*program counter*) value, containing the address of the next instruction byte to be fetched from EPROM, will then be used to drive the 16 address lines.

2) The lower eight address lines A0..7 are driven on Port 0 (also known as AD0..7 since it is multiplexed with A0..7 and D0..7), at the same time as A8..15 are driven on Port 2. At this point, the complete 16-bit address of the next instruction is available on Port 2 and the address latch. As soon as the address lines are stable and valid, the address access time for the memory begins. Since the address will be valid before ALE goes low, a transparent latch is used to give the memory the address as soon as possible.

If a negative edge triggered register was used instead of a transparent latch to hold the address bits, then address bits A0..7 would not be available until a propagation time after the falling edge of ALE.

3) Once the address lines are valid, ALE goes low, latching A0..7 bits in the external address latch. This allows the multiplexed data lines to be used for data transfer without disturbing the lower eight address lines that are held in the latch for the remainder of the cycle. Since the upper eight address lines (A8..15) are not multiplexed, they remain valid for the rest of the cycle and do not need to be latched.

4) /PSEN goes active (low) to indicate that this is a program memory read cycle, enabling the EPROM to drive the data bus. This enable signal begins the program memory read access cycle time for the EPROM.

5) /PSEN goes inactive (high) signaling the end of the program read cycle and clocking the EPROM data into the processor. Because this signal is used to clock data into the CPU from the data bus, there are associated setup and hold times for the data relative to the rising edge of the /PSEN signal.

Using the preliminary design, the first parameters to be investigated are the access times. Table 6-1 gives the program memory timing parameters for the 8031. The memories that have longer access times are less expensive than the fast ones, so we would like to use the least expensive parts that will meet the specifications. Both the address and enable access times are of interest, including all possible propagation paths for these signals. The slowest path will determine the maximum clock frequency that can be used for reliable operation, up to 12 MHz, the maximum CPU clock frequency. The ALE path will be ignored for now. All three paths must be evaluated to determine which one is the speed limiting condition. The three signal propagation paths for the program read cycle are:

a) Valid address A8..15 on Port 2, EPROM address access time

b) Valid address on port 0, D to Q delay through the latch, and EPROM address access

c) /PSEN active, EPROM enable access time

These three propagation paths are shown in Figure 6-3. Figure 6-4 shows the program memory timing diagram for the 8031.

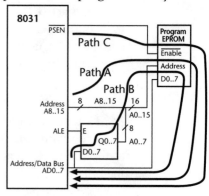

Figure 6-3: Three access propagation paths for program read.

A Detailed Design Example

Symbol	Parameter	12 MHz Clock min	max	units	Variable Clock 1/TCLCL = 1.2 to 12 MHz min	max	units
TCLCL	Oscillator Period	83		nS	83	833	nS
TCY	Minimum Instruction Time	1.0		uS	12TCLCL		nS
TLHLL	ALE Pulse Width	140		nS	2TCLCL-30		nS
TAVLL	Address Set Up to ALE	60		nS	TCLCL-25		nS
TLLAX	Address Hold After ALE	50		nS	TCLCL-35		nS
TPLPH	/PSEN Width	230		nS	3TCLCL-20		nS
TLHLH	/PSEN, ALE Cycle Time	500		nS	6TCLCL		nS
TPLIV	/PSEN to Valid Data In		150	nS		3TCLCL-100	nS
TPHDX	Input Data Hold After /PSEN	0		nS	0		nS
TPHDZ	Input Data Float After /PSEN		75	nS		TCLCL-10	nS
TAVIV	Address to Valid Data In		320	nS		5TCLCL-100	nS
TAZPL	Address Float to /PSEN	0		nS	0		nS

NOTE: Test Conditions T=0–70°C, Vcc= 5V±5%   Port 0, ALE and /PSEN Outputs: C_L = 150 pF
All Other Outputs: C_L = 80 pF

Table 6-1: 8031 program memory timing parameters.

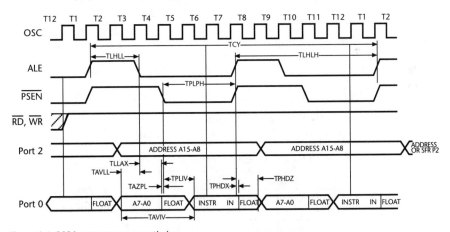

Figure 6-4: 8031 program memory timing.

Assuming a 12MHz clock, the timing analysis for the three paths is:

Path A

The delay from when the CPU provides a valid address A8..15 onPport 2 until
the end of the EPROM address access time, resulting in valid data from the
EPROM on the data bus. Figure 6-5 shows the EPROM timing diagram. The

CPU requires that the data from the EPROM be available 320 nanoseconds (nS) (TAVIV) after being presented with a valid address. The -30 version of the EPROM has an address access time of 300 nS max. (EPROM t_{AA}), so there is 20 nS of margin for this EPROM at this clock speed.

TAVIV - EPROM t_{AA} = 320 - 300 = 20 nS margin

Parameter	Symbol	Test Conditions	-15 min	-15 max	-20 min	-20 max	-25 min	-25 max	-30 min	-30 max	Units
Address access	t_{AA}	/CE=/OE= V_{IL}		170		200		250		300	nS
/CE access	t_{CE}	/OE= V_{IL}		170		200		250		300	nS
/OE access	t_{OE}	/CE= V_{IL}	10	60	10	70	10	100	10	120	nS
Output disable	t_{DF}	/CE= V_{IL}	0	50	0	50	0	60	0	105	nS

Table 6-2: EPROM timing parameters.

Note that the -15, -20, etc. at the top of Table 6-2 are suffixes that refer to the memory access times.

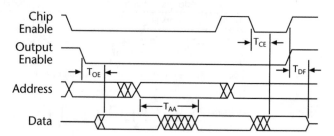

Figure 6-5: EPROM timing diagram.

The CPU ALE line is connected directly to the latch enable input of the 74ALS373 transparent latch. Table 6-3 gives timing specifications for this device. Remember that this type of latch simply passes the D inputs directly through to the Q outputs (after a propagation delay), as long as the enable input remains high. The '373 type latch has an *asymmetrical* propagation delay from the D input to the Q output, since t_{PLH} from D->Q is 12 nS max, and t_{PHL} is 16 nS max. This corresponds to the first part of the propagation path B.

As can be seen in Figure 6-2, ALE goes high before the address goes valid. The delay from the enable (E) input to

Parameter	From (input)	To (output)	min	max	Unit
t_{PLH}	D	Q	2	12	nS
t_{PHL}	D	Q	4	16	nS
t_{PLH}	E	Any Q	6	22	nS
t_{PHL}	E	Any Q	7	23	nS
t_{PZH}	/OC	Any Q	6	18	nS
t_{PZL}	/OC	Any Q	5	20	nS
t_{PHZ}	/OC	Any Q	2	10	nS
t_{PLZ}	/OC	Any Q	2	12	nS

Table 6-3: Timing specifications for 74ALS373 transparent latch.

the output (Q) is only 23 nS, much less than the time the CPU takes to put its address out on the bus. The delay in the ALE path is:

TLHLL-TAVLL = 140 - 60 = 80 nS

Since the latch is enabled in 23 nS, but the address is not available from the CPU until 57 nS later, this path is not considered. In this, as in most designs, the ALE delay path is not critical, so it is ignored. This must be considered for some CPUs, such as the Dallas Semiconductor high-speed 80C320 family of microcontrollers. Path B, from D to Q, is always worth examining.

Path B

From the time a valid address is available on port 0 (the multiplexed bus), plus the maximum D to Q delay through the latch, and the EPROM address access time, until valid data is on the bus.

The CPU allows the same total of 320 nS delay time for this path as above. In this case however, there is the additional delay of the latch that reduces the time available to the memory. The latch is specified for a maximum D to Q delay, $t_{P D->Q}$ of 16nS worst case. So from the 320 nS available, 16 nS is used by the latch, and 300 nS is used by the EPROM, leaving only four nanoseconds of margin!

$$\text{TAVIV - EPROM } t_{ACC} \text{ - Latch } t_{P D->Q} = 320 - 300 - 16 = 4 \text{ nS margin}$$

This is a slim, but acceptable margin, as long as the device outputs can drive the actual loads on their outputs. If the load capacitance exceeded the specified test load capacitance usually listed in the notes in the timing section, then the rise/fall time would be extended, possibly throwing this design out of the specified limits at the full 12 MHz clock speed.

Path C

For Path C, we need to evaluate the delay between the time the CPU enables the program memory and when the memory instruction output appears on the bus. The enable access time is from the activation of /PSEN, which enables the EPROM chip enable (/CE), until the EPROM provides a stable and valid instruction on the data bus.

Once again, the design margin is the time allowed by the CPU, less the time taken by the external circuits. The CPU allows TPLIV or 150 nS.

TPLIV - EPROM t_{CE} = 150 - 300 = -150 nS NEGATIVE design margin!

When /PSEN is directly connected to the EPROM /CE line, the CPU provides 150 nS (TPLIV) for the EPROM enable access time, but the -30 EPROM t_{CE} is 300 nS, which is *150 nS TOO SLOW!*

At this point, we have several options:

• Decrease the CPU clock speed.

• Buy a faster EPROM.

• Change the wiring: connect /PSEN to /OE instead of /CE.

Let's examine these three alternatives more closely.

1) **Reduce the clock speed of the CPU to conform to the EPROM's chip enable access time.** This has the obvious disadvantage that the processor will run more slowly.

2) **Buy an EPROM with faster chip enable access time.** Faster parts cost more and, in this case, the fastest device in the table has a chip enable access time of 170 nS, which is *still* too slow.

3) **Rewire the /PSEN line to EPROM output enable input (/OE) and connect the chip enable (/CE) to ground.** This does not require slowing the chip or using a faster, more expensive memory.

There is one other solution that is not available on the standard 8051 processor: the use of "wait states" which stretch the memory cycle timing by one or more clock cycles. The standard 8051 family parts do not incorporate this feature, but the high-speed versions do. The 80C320 family of high-speed microcontrollers from Dallas Semiconductor does allow wait states. These devices have internal registers that can be programmed to stretch memory cycles as needed to accommodate slower memories. Some other types of processors require external hardware to insert wait states.

Comparing all options, the simplest solution is probably 3). Let's see what happens to the Path C timing design margin calculation when we use that approach. In this version, the CPU's /PSEN line drives the EPROM's /OE input, with the /CE grounded. As before, the CPU allows TPLIV or 150 nS, but in this case we use the EPROM's output enable access time, t_{OE}. Looking back at the EPROM specifications, we find that for the slowest (-30) part, the worst-case value for t_{OE} is 120 nS.

TPLIV - EPROM t_{OE} = 150 - 120 = +30 nS design margin

When /PSEN is directly connected to the EPROM /OE line, the CPU provides 150 nS (TPLIV) for the EPROM enable access time, and the -30 EPROM t_{OE} is 120 nS, which is more than fast enough. This design change allows the CPU to run at the full 12MHz rating. The example shows how we may have to change the design in order to optimize the timing, and the iterative nature of the design process.

As in everything else, there are some drawbacks and implications for this approach that need to be considered:

- **The EPROM is always enabled when the /CE input is grounded, so only one EPROM can be used this way.** This has the disadvantage that the EPROM draws its maximum operating power constantly.

- **Use of /CE to enable the device reduces power consumption, which is important for battery powered applications, especially when there are multiple devices.** Enabling with the /CE input allows for the use of multiple memory chips in the system by using a memory address decoder to decode the appropriate address range. The decoder output can drive the selected memory device /CE input lines one at a time, just as we saw in the previous module on memory address decoding. That way only one of the memory devices is powered at a given time. The memories' /OE lines would be connected to the processor's /PSEN signal output, so that slower memories could still be used. As is the case for other specs, the speed or power consumption of the system can be optimized.

This concludes our example, but it is evident that there are many other timing specifications that must be evaluated for a given design. Fortunately, the same methods we have used here are applicable to the other timing specifications and devices used in a typical embedded controller system. This completes the preliminary evaluation of the program fetch cycle memory access times, which are often among the most difficult to meet. The next step is to analyze the data memory cycle timing.

External Data Memory Cycles

Data memory read and write cycles are also examined in basically the same way, using the CPU data read cycle data and the SRAM performance specifications. The data read cycle has essentially the same three possible paths as the program read cycle, except that the CPU /RD signal is connected to the SRAM /OE input, and the SRAM chip enable is grounded.

External Memory Data Memory Read

The data memory cycle corresponds closely to the program memory cycle, as shown in the accompanying figures and tables. Figure 6-6 illustrates the timing relationship between the CPU and external SRAM data memory when the CPU

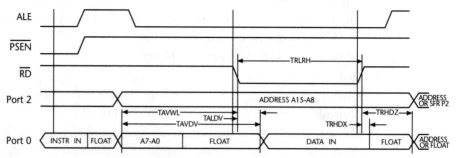

Figure 6-6: 8031 data memory read timing.

		12 MHz Clock			Variable Clock 1/TCLCL = 1.2 to 12 MHz		
Symbol	Parameter	min	max	units	min	max	units
TRLRH	/RD Pulse Width	400		nS	6TCLCL-100		nS
TWLWH	/WR Pulse Width	400		nS	6TCLCL-100		nS
TRLDV	/RD To Valid Data In		250	nS		5TCLCL-170	nS
TRHDX	Data Hold After /RD	0		nS	0		nS
TRHDZ	Data Float After /RD		100	nS		2TCLCL-70	nS
TAVDV	Address to Valid Data In		600	nS		9TCLCL-150	nS
TAVWL	Addressto /WR or /RD	200		nS	4TCLCL-130		nS
TQVWH	Data Setup Before /WR	400		nS	7TCLCL-180		nS
TWHQX	Data Held After /WR	80		nS	2TCLCL-90		nS

NOTE: There are 2 to 8 ALE cycles per instruction. Clocks and state timing are shown on the timing diagram for reference purposes only. They are not accessible outside the package. TCY is the minimum instruction cycle time that consists of 12 oscillator clocks or two ALE cycles. Address setup and hold times are the same for data and program memory.

Table 6-4: 8031 data memory timing parameters.

reads from the SRAM while Figure 6-7 shows the SRAM read cycle timing diagram. Table 6-4 gives the data memory timing parameters for the 8031, and Table 6-5 lists the SRAM's ready cycle timing parameters. The CPU's TAVDV spec places an upper limit on the data memory's access time, t_{AA}, for path A.

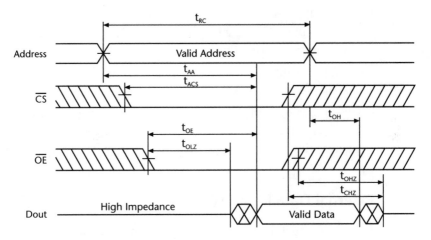

Figure 6-7: SRAM read cycle timing diagram.

Parameter	Symbol	-8 min	-8 max	-10 min	-10 max	-12 min	-12 max	-15 min	-15 max	Units
Read Cycle	t_{RC}	85		100		120		150		nS
Address access	t_{AA}		85		100		120		150	nS
/CS access	t_{ACS}		85		100		120		150	nS
/OE to Output Valid	t_{OE}		45		50		60		70	nS
Output hold from addr	t_{OH}	5		10		10		10		nS
/CS to output enable(low Z)	t_{CLZ}	10		10		10		10		nS
/OE to output enable(low Z)	t_{OLZ}	5		5		5		5		nS
/CS hi to out disable(hi Z)	t_{CHZ}	0	30	0	35	0	40	0	50	nS
/OE hi to out disable(hi Z)	t_{OHZ}	0	30	0	35	0	40	0	50	nS

Table 6-5: SRAM read cycle timing parameters.

A) The delay from when the CPU provides a valid address A8..15 on Port 2 until the end of the SRAM address access time, resulting in valid data from the SRAM on the data bus. The CPU requires that the data from the SRAM be available 600 nS (TAVDV) after being presented with a valid address. The -15 version of the SRAM has an address access time of 150 nS max. (SRAM t_{AA}), so there is 450 nS of margin for this memory at this clock speed!

$$TAVDV - SRAM\ t_{AA} = 600 - 150 = 450\ nS\ margin$$

B) Even allowing for an additional 16 nS through the address latch for address bits 0..7, there is still a margin of 434 nS, so there is no problem with address access time.

$$TAVDV - SRAM\ t_{AA} - Latch\ t_{Pmax} = 600 - 150 - 16 = 434\ nS\ margin$$

C) This is the time available to the memory after /RD goes low and when valid data is on the bus. The enable access time provided by the CPU is 250 nS (TRLDV). Since the slowest RAM, the -15 version, has an OE access time of 70 nS (t_{OE}), there is 180 nS of design margin.

External Data Memory Write

Figure 6-8 and Table 6-6 show the SRAM write cycle diagram and timing parameters. Figure 6-9 shows a data memory write timing diagram for the 8031.

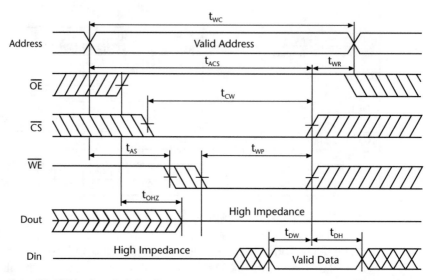

Figure 6-8: SRAM write cycle timing diagram.

Parameter	Symbol	-8 min	-8 max	-10 min	-10 max	-12 min	-12 max	-15 min	-15 max	Units
Write Cycle	t_{WC}	85		100		120		150		nS
Chip Select to end of write	t_{CW}	75		80		85		100		nS
Addr valid to end of write	t_{AW}	75		80		85		100		nS
Address setup time	t_{AS}	0		0		0		0		nS
Write Pulse width	t_{WP}	60		60		70		90		nS
Write recovery time	t_{WR}	10		0		0		0		nS
Write to output in high Z	t_{WHZ}	0	30	0	35	0	40	0	50	nS
Data to Write time overlap	t_{DW}	40		40		50		60		nS
Data hold from write time	t_{DH}	0		0		0		0		nS
Output disable to out in highZ	t_{OHZ}	0	30	0	35	0	40	0	50	nS
Output active from end of WR	t_{OW}	5		5		5		5		nS

Table 6-6: SRAM write cycle.

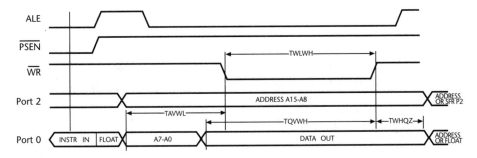

Figure 6-9: 8031 data memory write timing.

From the CPU specifications, the address is valid 200 nS (TAVWL) before the /WR line goes low, and the data is valid 400 nS (TQVWH) before the /WR line goes high. The RAM requires an address setup before write time of 0 nS, which is compatible with the 200 nS provided by the CPU. The RAM data setup time before the end of the /WE pulse (SRAM spec t_{DW}) is 60 nS, which is well within the 400 nS available. The latch delay has been ignored here because it is 16 nS, which is insignificant compared to the design margin available. Also, the chip select input of the RAM is grounded, so the chip select access time does not need to be considered. The minimum write pulse width from the CPU is 400 nS (TWLWH), and the RAM requires only a minimum of 90 nS (t_{WP}), so the pulse width is well within the spec. The RAM has a 0 nS hold time requirement (t_{DH}), and the processor provides 80 nS (TWHQX), so the RAM hold time requirement is also met with margin.

We'll now look at three typical design problems and show how to use the techniques described in this chapter to solve them.

Design Problem 1

For the same three paths in Figure 6-3, find the maximum allowable clock rate, given the slowest EPROM from Table 6-2. Use the specs for the -30 part which has a 300 nS access time and the same address latch specs in Table 6-3. Consider the 8031, EPROM, and 74ALS373 latch specs as discussed in the sections describing Paths A, B and C.

Solution: In this case, we are given the component timing, and we need to solve for the minimum clock period (T = 1/maximum clock frequency).

Path A:

The CPU allows TAVIV = 5*T-100 nS
The EPROM uses Taa = 300 nS
The limiting condition is TAVIV = Taa, so:

> 5T-100 = 300
> 5T = 400
> T = 80 nS

Path B:

The CPU allows TAVIV = 5*T-100 nS
The EPROM uses Taa = 300 nS
The latch uses TPHL D->Q = 16 nS
The limiting condition is Taa + Tlatch = TAVIV, so:
TAVIV = Taa + Tplatch and TAVIV = 5T-100, so:

> 5T-100 = 300 + 16
> 5T = 416
> T = 83 nS

Path C:

The limiting condition is TPLIV = Toe of the EPROM, so:
The EPROM Toe from the table is 120 nS
The equation is TPLIV = Toe

Solving for T, we have:

$$3T - 100 = 120$$
$$3T = 220$$
$$T = 220/3 = 73 \text{ nS}$$

Of all three paths, the longest period is due to Path B at 83nS, so it is the limit to the clock rate for the specs considered here.

Paths A and C are not constraints for this case.

So Path B is the limiting case when /OE is connected to /PSEN, and the maximum clock frequency is 1/83nS = 12 MHz.

Note that Path B is just at the spec limit for 12 MHz operation (1/83 nS = 12 MHz), so the maximum clock is 12 MHz, even for a faster EPROM.

Also notice that if /PSEN was instead connected to /CE, (Path C), the TPLIV spec would be the limiting factor: TPLIV = 3T-100 = Tce of the EPROM. The EPROM Tce from the table is 300 nS. Solving for T, we have:

$$3T-100 = 300$$
$$3T = 400$$
$$T = 400/3 = 133 \text{ nS.}$$

For this case, 1/133 nS = 7.5 MHz would be the maximum allowable clock rate.

Design Problem 2

You have an existing processor design, and you need to define what the minimum acceptable specs are for the program EPROM to determine which vendors and part numbers will work in the system. Assuming a clock rate of 12 MHz for the 8051, determine the following specs for the memory chip to be used with it, assuming the same address latch used in the previous examples, and find the maximum acceptable values for:

- Tce max (chip enable acess time)
- Taa max (address access time)
- Tod max (output disable time, referred to as Tdf in the EPROM spec)

Assume /PSEN is connected to the EPROM /CE and EPROM /OE is grounded.

Solution: In order to determine the required Tce, we need to calculate the memory spec based on the CPU speed. Since /PSEN is connected to the EPROM /CE, the relevant CPU spec is TPLIV. From the 8031 program memory timing table, TPLIV = 3T-100 nS, where T is TCLCL, the clock period. The answer for Tce is in the table for 12 MHz as 150 nS, but it could be computed for an arbitrary clock as:

Tce max = 3*83.3-100 = 150 nS

Taa is different, because the latch delay must be included. In this case the relevant CPU spec is TAVIV, which is 320 nS at 12 MHz. Subtracting the worst case latch delay, Tphl D->Q is 16 nS. Therefore only 320-16 = 304 nS is available to the memory as Taa. The general solution is TAVIV = 5T - 100, so:

Taa = TAVIV-Tplatch = 5*83.3-100-16 = 301 nS

Note that the Taa result is slightly (3nS) different from the value computed using the table. This is not unusual because the specs are not necessarily consistent, nor are they precise to a few nS. Many of the specs are based on statistical estimates of the production population, and are themselves only approximations. Often these specifications are guaranteed but not tested on every device.

Tdf is the time the EPROM takes to turn off its output drivers. This relates to the time the CPU allows for the EPROM to turn off its tri-state driver outputs after /PSEN goes inactive. If this spec is violated there will be bus contention between the CPU and the EPROM for the time of the overlap. The relevant CPU spec is TPHDZ. At 12 MHz, 75 nS are available to the EPROM to disable its outputs. The general form is TPHDZ = T-10 or 73 nS, again slightly different from the table value.

Design Problem 3

For a specific EPROM spec, find the maximum allowable clock rate, given the slowest EPROM from Table 6-2. Use the specs for the -30 part which has a 300 nS access time and the same address latch. Consider the 8051 specs for TPLIV, TAVIV, and TPHDZ.

Solution: In this case, we are given the component timing, and we need to solve for the minimum clock period (= 1/maximum clock frequency).

The equation for TPLIV = 3T-100 = Tce of the EPROM. The EPROM Tce from the table is 300 nS. Solving for T, we have:

3T - 100 = 300
3T = 400
T = 400/3 = 133 nS.

The EPROM Taa = 300 nS, Taa = TAVIV-Tplatch and TAVIV=5T-100, so:

300 = 5T - 100 - 16
5T = 416
T = 83 nS

The EPROM Tdf = 105 nS, and Tdf = TPHDZ = T-10, so:

105 = T - 10
T = 115 nS

Of all three specs, the longest period is due to the EPROM Tce and TPLIV spec, 1/133 nS = 7.5 MHz. If the /PSEN signal is connected to the -30 EPROM's /OE pin however, then:

EPROM Toe = 120 nS
TPLIV = 3T - 100 = 120
3T = 220
T = 73 nS

With /PSEN connected to /OE the TPLIV spec is not the limit.

The next slowest is due to TPHDZ, resulting in a minimum clock period of 115 nS, corresponding to a maximum clock frequency of 1/115nS = 8.696 MHz.

For the -30 EPROM in Table 6-2, Tdf will be the limiting specification when the access time is fast enough. Tdf is 105 nS, which is greater than the 75 nS available at 12 MHz, resulting in as much as 105 - 75 = 30 nS of bus contention! That is a serious conflict, and should not be allowed to occur.

Note that the access times were well within specifications for 12 MHz operation. If we looked only at the access time specs there is no problem, so the system might appear to work. However, bus contention may occur at the 12 MHz frequency, so the correct answer is 8.7 MHz.

If we change the EPROM to the -25 version, it is possible to clock the CPU at its limit of 12 MHz without exceeding any of the other specs. This example shows why it is important to consider ALL the specs, since it is not always the obvious specs that are the limits.

Completing the Analysis

Once the preliminary timing analysis is complete, the next step is to evaluate the noise margin as well as the DC and AC loading for the design. The results of this will determine if any of the signals are incompatible or overloaded, requiring changes to the circuit design or component selection. Of course, any changes made to the design (changing components, adding pull-up resistors, etc.), will require the timing to be re-evaluated. Once again we find that the interactions may cause us to do our design in an iterative fashion. This is part of the reason we don't want to perform a complete timing analysis from the beginning.

Once the preliminary timing, noise margin, and loading analyses indicate that the design is correct, it is necessary to review *all* the remaining specifications for *all* the ICs used in the design. This is not as difficult as it might seem. Most of the hard work is done as part of the preliminary analysis. Also, many of the device specs are simply not applicable to a given design. Examples of these specs include alternative SRAM memory write cycles. A given processor will always use one particular memory write sequence (i.e.: address stable first, then /CS active, then /WE goes low). As a result, the other write cycles and specs can be ignored. Still other specifications are just for information, such as the 8051 TCY spec, which simply informs us that an instruction cycle takes 12 clock cycles on the standard 8051. There will be some other specs that will apply to our design, such as the setup and hold times for some devices. In some cases the specification is a non-constraint, such as the 8051's TPHDX, input instruction data hold time after /PSEN goes high, specified as 0 nS. A zero hold time indicates that the driving device may remove the instruction at the instant when /PSEN goes inactive. Any device will meet that constraint, since it cannot predict in advance when the /PSEN line will change. Other specs will often have a huge margin as can be seen by inspection. The '74ALS373 address latch, for instance, requires a minimum enable pulse width that is on the order of 10 nS. The CPU puts out an ALE pulse that is TLHLL = 140 nS wide, so there is obviously lots of margin in that case.

With experience, this iterative design and analysis process becomes much easier, and potential problems are easier to anticipate. However, even with experience it is easy to become lax and leave out the review of the seemingly less important specs. This will often result in a direct application of Ken's first law of worst-case analysis: *"Any specification which is not considered will certainly be violated, causing catastrophic failure at the worst possible time."* That's usually right before a salary review or in front of an important customer! It is important to review all the specs for the parts to be used in a design. When alternate sources for the devices are to be used, the specifications of these alternates should also be reviewed. Parts from two vendors with the exact same part number may have subtly different specs.

Chapter Six Problems

1) For this problem, use the fastest EPROM program memory from Table 6-2 (the –15 version), the 8031 CPU specs in Table 6-1, and the latch specs from Table 6-3. Ignoring the TCLCL limit on clock speed, how fast can the processor be clocked? Use the connections shown in Figure 6-3, with /PSEN connected to the EPROM /CE pin.

2) Use the same conditions as the problem above, except connect /PSEN to the EPROM /OE control.

3) For a system that has multiple program memories, an address decoder is required in order to generate separate select signals to enable the program memories. What paths and specs will be affected and how will the timing change?

4) For each of the CPU data memory write timing parameters listed in Table 6-5, list the corresponding SRAM timing parameters from Table 6-6.

Programmable Logic Devices

Application specific integrated circuits (ASIC are ICs that have been designed or programmed to meet the needs of a specific design in which the chips will be used. These are differentiated from standard, or general purpose ICs that may be used in many different applications. General-purpose logic ICs are usually designed "from scratch" using only the most basic circuit elements such as transistors and gates. The cost of building a chip this way can be amortized over a large number of devices if it is used in many different applications.

When an application specific chip is designed from scratch, it is referred to as a *full custom* logic design. It is the lowest cost to manufacture because it takes the least amount of silicon to implement a given function. Unfortunately the design of a large full custom chip is very expensive (hundreds of thousands to millions of dollars) due to the labor-intensive design and prototyping process, and cannot be justified unless a very large quantity will be manufactured. Originally this was the only way to design chips, but now there are several alternatives for designing ASICs.

Standard cell IC design uses a library of common logic functions that have already been designed and tested. This reduces the amount of design effort in that logic IC blocks such as multiplexers are used in place of the equivalent random logic design implemented with gates. The cells can range in complexity from simple gates to complete CPUs. Standard cell based IC design has become the standard and can now be done even on a PC at a much lower cost than other methods. The cost of manufacturing a minimum production quantity of parts is less (thousands of dollars) than it would be for a full custom design process, but still high enough to be inappropriate for prototyping and low volume production (e.g., less than 5000 units).

Gate arrays are fabricated with a fixed array of gates and the wiring is defined by the user when the chip is manufactured. The advantage of this approach is that the chips can be fabricated up to the point where the interconnecting wires are placed on the chip, ready for a custom interconnect. Since most of the processing is completed before the connections for any particular design are placed on the chip, these "almost finished" devices can be produced and stockpiled without any interconnects. The final interconnect can be added to define a given design, resulting in a customized version of a nearly standard part. Gate arrays are often used when the production volume is too low to justify a full custom design, but high enough so that a user programmable device is not cost effective.

User *programmable logic devices* (PLDs) are a family of devices that are all manufactured in the same way, and can be customized using a special pro-gramming process much like an EPROM. An EPROM can be used to implement arbitrary logic functions by using the address lines as inputs and data lines as outputs. Thus, a 1Mx8 EPROM could have eight independent logic outputs that can be *any* boolean function of any of the 20 input address bits. The fully completed truth table is programmed into the EPROM so that each unique input pattern will result in the appropriate data at the outputs. EPROMs are not often the best choice for the kind of logic required in most designs because of their speed and relative complexity, which translates to performance and price. Other types of devices have been designed specifically for use in those appli-cations. There is a wide range of devices available, from fuse-linked two-level combinatorial logic and EPROM registered logic arrays, to arrays of logic blocks programmed with SRAM memory in each block. These devices span a range of complexity from one hundred to more than ten thousand usable gates. These families of devices are referred to as *programmable logic arrays* (PLA), *programmable array logic* (PAL, a registered trademark of AMD Inc.), and other trademarked names.

Field programmable gate arrays (FPGAs) are a cross between gate arrays and PLDs. They have an array of logic with user programmable interconnections. FPGAs are generally used where the desired logic function is too large to fit in a sum-of-products device, and the volume is too low to justify the use of a gate array or custom logic. FPGAs are available in sizes large enough to implement an entire CPU.

Introduction to Programmable Logic

The most common types of programmable logic are two level (AND-OR) logic chips implementing a sum-of-products logic function on each output.

An example sum of products form in standard notation is: **F = AB + CDE**

The notation used in this book for the example above is: **F = A*B + C*D*E.** The conventions we will follow include:

- Logic AND is denoted by an asterisk: *
- Logic OR is denoted by a plus sign: +
- Logic inversion (NOT) by a slash: /

The examples above would require three gates: one two-input AND gate, one three-input AND gate and a two- input OR gate to combine the AND gates' outputs. (Other references may use different notation, such as & for logic AND, or a minus sign - for inversion, e.g.: **F = A & B + -C.**)

There are several varieties of two level programmable logic devices, with most of the variations relating to the type of output. Some devices have output flip-flops to allow storage and sequential logic, and some have tri-state drivers. The outputs of some of these devices can be defined at the time they are programmed as inverting, non-inverting, latched, bi-directional, asynchronous and other configurations. The pattern used to program the device is referred to as a *fuse map* because the original chips used fuse linked memory and the map represents the pattern of blown fuses.

Technologies: Fuse-Link, EPROM, EEPROM, and RAM Storage

Fuse-link PLDs consist of an array of fuses that make connections between the inputs and the logic gates inside the chip. When the chip is programmed, the unwanted fuses are "blown open" to leave only the desired connections. Fuse-link devices are implemented using bipolar logic so they are very fast, and consume a lot of power. Obviously they can only be used once, so they are not as desirable for prototyping purposes as an erasable device. Erasable parts, built using the same technology as EPROM, EEPROM, and RAM data storage for the arrays are available and carry with them the same characteristic advantages and disadvantages as their respective memory types.

Architectures

The first user programmable logic array chips had two levels of asynchronous logic. They were organized with two arrays of programmable fuse links, one connecting the inputs to an array of AND gates and the other connecting the AND gate outputs to an array of OR gates driving the output pins. This type of device allows arbitrary sum-of-products logic functions to be implemented limited only by the number of AND and OR gate inputs, and the I/O pins.

Programmable array logic devices are similar to PLA devices except that there is only one fuse array connecting the inputs to the AND gate array. The connections between the AND and OR gates in the PAL are fixed by the design of the PAL. Both PLAs and PALs are made with either active high or active low outputs. It is important to note that the arrays and inputs are not necessarily identical; some OR gates in a PAL may have more inputs than others on the PAL, for example.

Field programmable gate arrays have a more general architecture, and are not limited to the sum-of-products form. FPGAs have programmable interconnecting wires, logic blocks, and I/O pins. The connections and logic in FPGAs are defined by use of either static RAM, E/EEPROM or *anti-fuses*. Anti-fuses are like fuses, except that they have a high resistance in the unprogrammed state and when programmed their resistance becomes much lower. The anti-fuse is programmed to make a connection by forcing a current through the anti-fuse. Anti-fuse FPGAs are based on an array of gates and wires that can be selectively shorted with the anti-fuse acting as a one time programmable short circuit. FPGAs are almost exclusively implemented in CMOS technology because of the high logic density to keep the chip power and temperature to reasonable levels. Static RAM based FPGAs are composed of logic blocks with embedded volatile static RAMs that must be loaded with configuration data every time they are powered on. The logic functions and interconnection information is stored in volatile static RAM. The configuration can be loaded from an EPROM or EEPROM directly or via a CPU before they are used.

Relatively large supply currents are drawn by bipolar PLDs, so CMOS versions have been made available to reduce the power consumption requirements. Most of the CMOS PLDs are actually mixed NMOS and CMOS logic, so their power dissipation is not as low as pure CMOS. Use of a PLD in a battery-powered application will generally require a pure CMOS PLD to maximize battery life.

Erasable (E/EE) versions are available from several vendors, which are particularly useful in the development and debug of a new design when things change frequently. The fuses are replaced with floating gate switches with essentially the same construction as the EPROM and EEPROM memory cells described earlier. The EPROM versions of these parts are sold in windowed packages so they can be erased just like a UV EPROM, as well as non-windowed packages that can only be programmed once (one-time programmable, or OTP). The EE versions of these parts are erased electrically before they are programmed.

Small programmable logic devices consist of an array of programmable connections, or fuses, interconnecting the input signals with a number of AND gates, followed by an array of connections between the AND gates and some OR gates, resulting in one or more "sum-of-products" logic outputs. The notation used to illustrate the fusible interconnections between the inputs, gates, and outputs is shown in Figure 7-1. Its purpose is to allow a compact pictorial representation of the circuits, by avoiding the explicit representation of each independent input signal to gates that have a large number of inputs. Instead of showing every gate input, a single line represents multiple inputs, and an "x" is placed at points where the gate inputs are connected to one of the PLD input signals. Figure 7-2 shows an example of this.

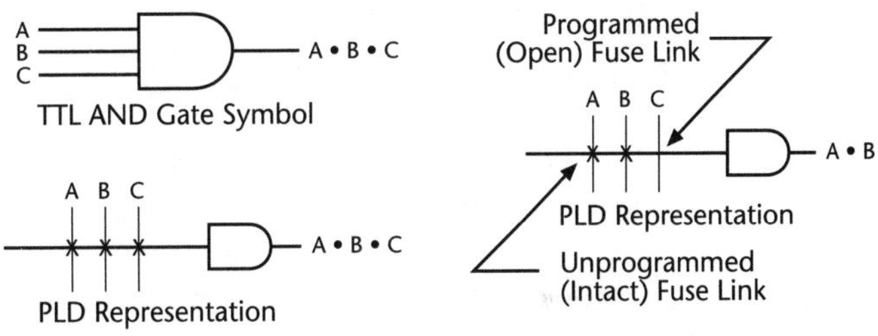

Figure 7-1: PAL logic diagram shorthand notation.

In understanding how various PLDs operate, it is useful to look at several ways in which a programmable logic device can

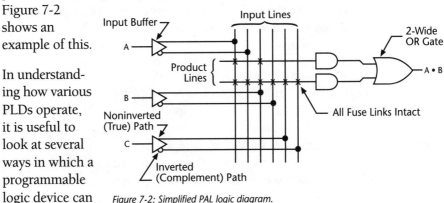

Figure 7-2: Simplified PAL logic diagram.

be organized. The simplest approach is to use a PROM memory as a programmable logic device, using the address lines as input and the data lines as output.

Figure 7-3 shows a PROM memory, with an array of AND gates connected to decode each memory address, and multiple bits per location that can be programmed by the user to output an arbitrary binary value (memory contents) for each combination of the inputs (addresses).

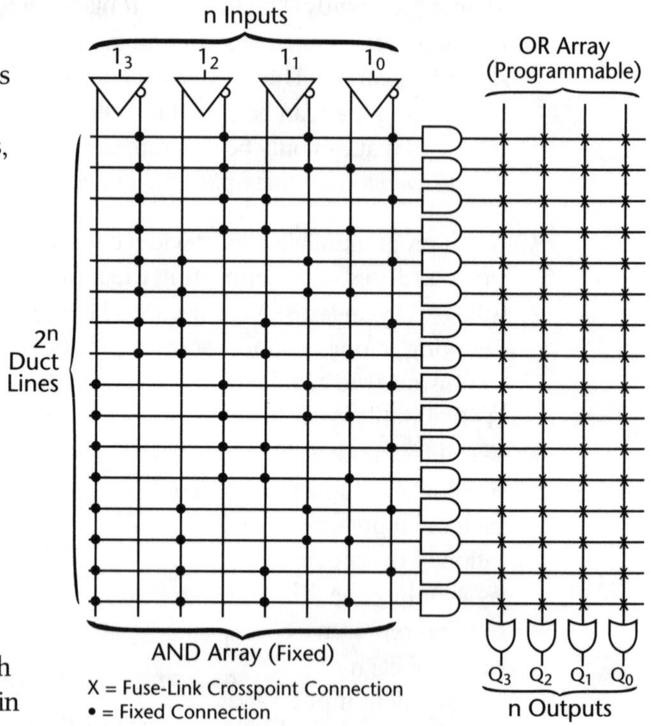

PROM as PLD

Figure 7-3 shows the fixed AND array which decodes each location in the PROM. Note the binary pattern of con-

Figure 7-3: Typical PROM as PLD architecture.

nections in the AND array. The top AND gate decodes address zero, enabling
the pattern programmed in the top row of fuses to be presented at the output.
This pattern is the 4-bit word of data stored in location zero as a pattern of
programmed fuses.

The advantage of using a PROM as a PLD is it can implement any logical
function of the inputs, regardless of complexity of the logic function to be
represented. This is because each possible permutation of the inputs corresponds to one memory location, and the PROM is essentially a physical
implementation of the complete logic truth table. Unfortunately, the number
of bits in the memory grows exponentially with the number of inputs. Since
most practical logic functions do not have very many product terms on
average, the memory is very sparsely filled with data. This means most of
the circuitry is effectively wasted.

Programmable Logic Arrays

The PLA is a very flexible logic device, as it allows both the AND as well as the OR arrays to be programmed by the user. Figure 7-4 illustrates the architecture of a typical PLA.

The PLA allows the implementation of almost any sum-of-products logic function to be implemented, within the constraints of the available number of input pins, AND gates, OR gates, and output pins. While the PLA architecture allows more efficient utilization of the resources on the chip, it is also more difficult to program, as fuses must be programmed in two separate arrays. Standard memory programming devices cannot be easily modified to program a PLA with two arrays.

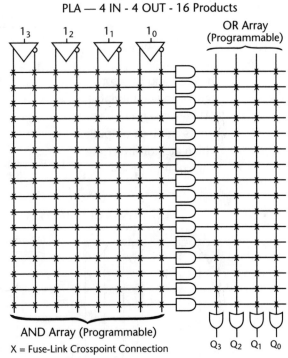

Figure 7-4: Typical PLA architecture.

PAL-Style PLDs

While there is a wide variety of programmable logic available, the most prevalent low cost version used in embedded designs is the PAL, a variation of the PLA sum-of-products chip. Consisting of a programmable AND (product) array and a factory-defined OR (sum) array, it is very similar to a standard memory device. As a result, many memory programmers can also be used to program PALs. This is a key reason for the success of these devices, along with the availability of software to ease in designing the fuse patterns for implementing specific users designs.

In a typical PAL, the inputs and their logical complements are provided to each of the AND gates through a programmable array of fuse connections.

The connections between the AND and OR gates are fixed by the manufacturer, and in most cases, some of the outputs are also fed back to the input array.

Figure 7-5 shows the PAL implementation of the logic function /(A * /B + /A * B).

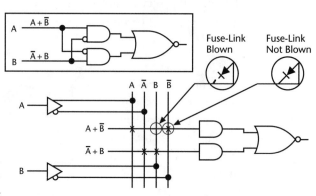

Figure 7-6 shows a simplified example of the logic and fuse configuration used in most PAL devices. It has four inputs and four outputs which are non-inverting sums of four products.

Figure 7-5: Example of PAL fuse programming.

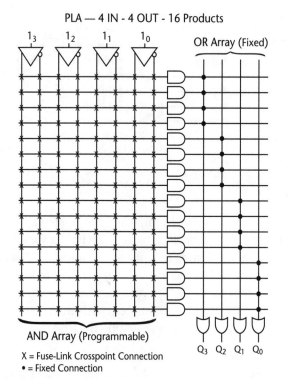

PLA — 4 IN - 4 OUT - 16 Products

OR Array (Fixed)

AND Array (Programmable)

X = Fuse-Link Crosspoint Connection
• = Fixed Connection

Figure 7-6: Typical PAL organization.

Most small PLD parts use a numbering convention that makes it easier to determine the configuration of the logic. The number is usually composed of three parts: the number of inputs to the array, output circuit type, and number of outputs. Thus a PAL with the part number 16L8 has 16 inputs to the AND array (not necessarily that many input pins), and eight active low outputs (L), while a 12H6 has 12 inputs, and six active high (H) outputs. A device number with an "R" in it has an output register, and a "V" indicates variable or user programmable outputs. Some of the pins may be shared inputs and outputs. Not all of the outputs are necessarily of the same type, however. The 16R4, for example, has four registered outputs and four asynchronous (un-registered) outputs. The V parts have a special output "macro-cell" that can be programmed to be asynchronous (un-clocked), synchronous (clocked), inverted, non-inverted, feedback internally to the AND array, and so on.

Design Examples

Probably the most common applications of simple PLDs are as address decoders in microcomputer and microprocessor systems. A device such as the 16L8 PAL, with active low outputs is well suited to drive the active low enable inputs of most memory and I/O devices. Because a PAL can have different logic functions on the same chip, one PAL can decode both memory and I/O addresses. Figure 7-7 shows an example of this. The program memory map for Figure 7-7 is shown in Table 7-1; the external data memory map is given in Table 7-2. Address and control lines can be wired to the input pins, and the output pins can drive the select and enable lines of the memory and I/O chips.

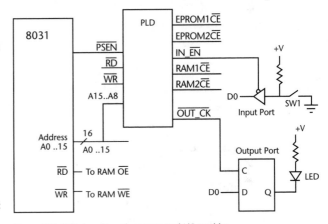

Figure 7-7: PLD decoding of memory and I/O enables.

Program Memory Address Space

Address Range (hex)	Device Selected
Program 0000 - 7FFF	2Kx8 EPROM 1
Program 8000 – FFFF	32Kx8 EPROM 2

Table 7-1: Program memory map for Figure 7-7.

External Data Memory Address Space

Address Range (hex)	Device Selected
Data 0000 - 7FFF	32Kx8 SRAM 1
Data 8000 - FEFF	32Kx8 SRAM 2
Data FF00 - FFFF Read	Input port enable
Data FF00 - FFFF Write	Output port latch clock

Table 7-2: External data memory map for Figure 7-7.

An 8031 system with two program EPROMs, two data SRAMs, and memory mapped I/O ports could be connected using gates or decoders, but it would be more efficient to use a 16L8 PAL. The inputs to the PAL are the control and address lines. The outputs are the memory chip enables, the input port drive enable, and the output latch clock.

Note that the program memory is full, using two 32-kilobyte devices. Also, the input and output ports appear in the external memory address space. This is an example partially decoded, memory mapped I/O, since the input and output devices appear repeatedly in a memory address range of FF00 to FFFF hex.

For the program memory, the equations that must be used to program the PLD would be as follows:

/EPROM1CE = /PSEN * /A15   Enabled when PSEN active and A15 = 0
/EPROM2CE = /PSEN * A15   Enabled when PSEN active and A15 = 1

The equations above will enable the EPROMs when the processor is fetching instructions (/PSEN = 0) so that EPROM1 will be enabled for program memory addresses 0-7FFF and EPROM2 will be enabled for addresses 8000-FFFF.

The gates in Figure 7-8 are the equivalent to the equations for the EPROM chip enable equations shown above. Note that the gate used for the EPROM1CE function is equivalent to a simple OR gate. This is because inverting all the inputs and outputs of a logic function changes it from an AND to an OR and vice versa.

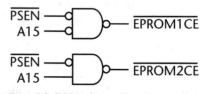

The RAM addresses are similarly

Figure 7-8: EPROM chip enable gate equivalents.

enabled when the processor is accessing external data memory, except that the input and output ports are memory mapped into addresses FF00 to FFFF. In order to avoid bus contention, SRAM 2 is disabled between FF00 and FFFF.

/RAM1CE = /A15   Enabled when A15 = 0
/RAM2CE = A15 * /A14 * /A13 * /A12 * /A11 * /A10 * /A9 * /A8
            Enabled when address > 8000 and < FFxx
/IN_EN = /RD * A15 * A14 * A13 * A12 * A11 * A10 * A9 * A8
            Input Enabled when RD low and address = FFxx
/OUT_CK = /WR * A15 * A14 * A13 * A12 * A11 * A10 * A9 * A8
            Output latch clock when WR address FFxx

This PLD implements a system with 64 kilobytes of program EPROM, and 64 kilobytes-256 bytes of data RAM. Note that 256 bytes of the external data memory address space are dedicated, or mapped, to the input/output port. The input port can be read as if it were data memory at locations FF00 to FFFF hex. This incomplete, or partial address decoding, decodes 256 different addresses to access the same input/output port (*partial I/O address decoding*). Similarly, the output port can be written to by writing data to data RAM addresses FF00 to FFFF hex. Note that changing the I/O addresses to different values requires only changing the equations and burning the corresponding fuse map patterns into another PLD. If the address map should need to be changed, it is possible to do so by using a different PAL device programmed with a different fuse map representing different equations.

Tables 7-1 and 7-2 above can also be represented graphically with an address map, as shown in Figure 7-9.

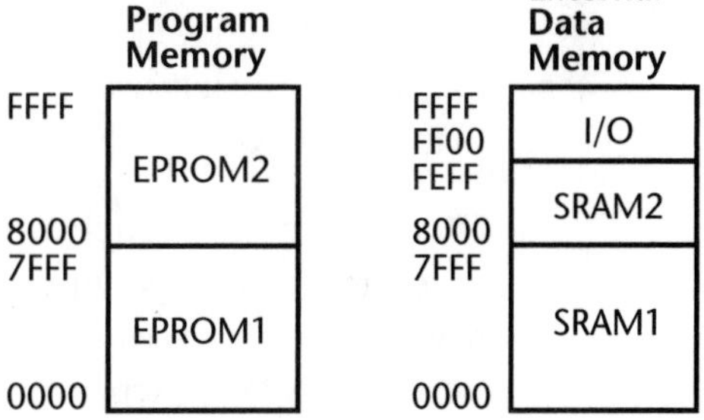

Figure 7-9: Memory address map.

PLD Development Tools

Because of the complexity that results from flexibility in defining PLD functions, it is not practical to manually define the fuse map that is used to program a PLD. Automated translation programs are used to convert a higher-level description of the logic to the low-level fuse map that is required. The software that performs that task is referred to as a *PLD assembler* or *compiler* because it is equivalent to a programming language translator used on a general-purpose desktop personal computer. PLD development software is available from both PLD vendors and other software houses, in versions that run on PCs and workstations. PLD assemblers input Boolean equations and generate the corresponding fuse map for programming. PLD compilers, on the other hand, take higher-level circuit descriptions such as logic schematics, state diagrams, and truth tables as input in addition to Boolean equations. The equation notation and syntax are unique to each particular translator. Some of the translators will perform additional functions such as selecting the appropriate type of PLD for the design, and logic minimization that is intended to reduce the complexity and cost of the device that will ultimately implement the design.

The two most common high-level logic compiler languages are VHDL and Verilog. Both of these hardware description languages are in common use for the design and definition of large, complex logic designs, as are commonly implemented in large custom logic ICs, and FPGAs. Because the larger FPGAs are difficult to program and modify using standard gate and module level design, the high-level hardware description languages are gaining popularity. The advantages and disadvantages of using a high-level hardware description language are very similar to those of a high-level computer language. By implementing a design using these high level descriptions, it is possible to take a chip design from one type of device to another with less effort than if it was done at the gate level. Of course, similar trade-offs exist as they do for high-level language programs. They are less compact and efficient in the way they utilize the hardware, and tend to result in somewhat slower performance than hand optimized gate level designs.

Other design tools, such as logic simulators, allow the logic functions to be tested against known input and output logic patterns. The test patterns, referred to as *test vectors*, are presented to the software simulation of the PLD logic design in sequence and the simulated outputs are compared with the desired outputs for discrepancies. The test vectors for design verification are generated by the design engineer to verify that the design will perform as intended. Unfortunately some of the most common problems and errors are those that were not planned for and only show up upon plugging the PLD into the circuit. Unforeseen conditions often cause erroneous outputs, requiring correction of the PLD design.

Test vectors are also important in verifying that the PLD is fully functional. Even though the fuse map is read back during the programming process, other faults may be impossible to detect by verifying the fuse map. This is particularly true for fuse link devices, since there is no way for them to be fully tested at the factory due to the fact that they cannot be erased. The test vectors that are used for design verification can be applied to the device for testing immediately after being programmed and verified on many PLD programmers. Generating a set of test vectors that will detect all possible faults (100% fault coverage is virtually impossible, and even approaching that goal can require a lot of effort and many test vectors, particularly for sequential circuits. To address that need, some PLD software vendors have *test vector generator* programs, which will create a set of test vectors for a given PLD design automatically. Most of the newer, more complex devices also have special test pins (JTAG, boundary scan) that improve the ability of a test system to modify and observe the state of the internal logic, which makes the tests easier and faster.

Simple I/O Decoding and Interfacing Using PLDs

Programmable logic is particularly useful for decoding the addresses and control lines from a processor, because it can be used to activate the chip enable signals for the various memories and I/O chips in a system. PLDs are more flexible than standard logic for several reasons. Each of the PLD outputs can be programmed to go active when the inputs are in a particular state, such as a particular address or range of addresses. The same functions that can be decoded in a PLD would generally take several standard logic chips. This is because many of the inputs, such as address lines, are common to several of the output logic functions. Also, because the devices are programmable, the decoding logic can be changed without changing the wiring on the printed circuit board. These characteristics have made PLDs very popular, which has in turn brought their prices down to levels that are comparable to standard logic solutions. The only disadvantage to using PLDs is that they require software to "compile" the logic into binary patterns and an instrument, equivalent to a PROM programmer, which can program the device with those patterns. Each type of device requires a special programming procedure, which may be unique to the manufacturer of the PLD. Generic compilers and programmers are available, but there are devices that can only be programmed using the manufacturers proprietary software or programmer.

IC Design Using PCs

For designs that must be very inexpensive in high volume, and for designs that must fit in a tight space, a custom logic IC may be the best solution. Custom ICs (ASICs) are also becoming easier to develop with the availability of PC-based IC design tools. Because PCs have become available to almost all design engineers, computer-aided design (CAD) software has been written to run on the PC for custom and standard cell IC design as well as PLD design. Some versions of this IC CAD software can be obtained for a few thousand dollars, making it practical even for smaller firms. Some versions of this low cost software will even convert from a schematic level circuit description to a detailed IC layout that can be transmitted via modem to an IC fabrication facility. In addition, MOSIS, a joint project of government and university organizations, has been operating for many years to provide low cost IC prototypes for the government, universities, and small companies who could not afford the high costs (many thousands of dollars) for a dedicated IC prototype run. By combining multiple designs on each silicon wafer, the minimum fabrication costs are reduced to as low as approximately $500 for a design with less than 1,000 gates, with delivery of six to eight weeks. This makes it practical for every engineer to design custom and standard cell ICs. Design oriented software is also available for simulation of the chip before prototyping begins. The logic functionality can be verified by implementing early prototype chips using PLDs. Production parts can then be made in low volume using MOSIS or by other vendors in high volume at lower cost. Advantages of this approach include fewer ICs, smaller size, lower power, control of proprietary designs, and lower cost in volume. An application requiring high levels of integration, low cost in high volume, or very small size would be most appropriate for this design approach.

FPGA devices allow the designer to prototype and change custom designs and test them quickly. Some of these devices store their logic configuration in SRAM memory, allowing the hardware to be re-programmed quickly, even in the final application. The largest devices contain the equivalent of about one million gates, and processors can easily fit on these larger chips along with a great deal of other circuitry. The 8051 CPU, can fit easily into one of the moderate size devices. Large building blocks or *IP cores* (IP = intellectual property) can be purchased from companies that specialize in their design. The core chip building blocks include CPUs, memories, I/O devices, data converters, and so on. These complex core building blocks can be combined on a single chip to achieve "systems-on-a-chip" (SOC). This is made possible because custom ASIC, and even FPGA devices, can accommodate a number of fairly complex core blocks on a single chip. Custom ASICs have large non-recurring expenses, but have the lowest cost in moderate to high volume. Large FPGAs are very expensive (often hundreds of dollars each) so the large devices may not be suitable for high volume applications unless they must be reprogrammable in the field. Some FPGA vendors are even promoting the idea that large SRAM based FPGAs could be updated through the Internet. This would allow the ultimate consumer to upgrade their hardware as easily as upgrading the software.

Chapter Seven Problems

1. How many pins would be required on a PLD in order to implement a completely decoded memory and I/O address decoder for the design shown in Figure 7-7?

2. For the problem above, make a revised version of Table 7-1, with the input and output ports mapped to address FFFF hex.

3. Write the two equations necessary to map the I/O port select signals, / IN_EN and /OUT_CK, of Figure 7-7 to respond only to address FFFF hex.

4. If a PROM is used to implement the PLD function above, how many memory bits would be required? How many fuses would be required of a PAL style version, using the PAL shown in Figure 7-6?

Basic I/O Interfaces

Ultimately computers are useless unless they are connected somehow to the outside world. This chapter emphasizes the connection of simple I/O (input/output) devices to a microcontroller, directly and mapped into the processor's memory or I/O address space using a bus. We'll also discuss more advanced I/O techniques.

For embedded processors, I/O capabilities are among the most important factors to consider when selecting a CPU. Typical microcontroller ICs have on-chip bi-directional parallel ports, serial ports, and timer/counter devices. Many also have specialized I/O for driving LCDs, analog-to-digital converters, pulse-width-modulated (PWM) digital-to-analog outputs, complex pulse trains of programmable width, and timers for period and frequency measurement, etc. Some devices also incorporate special serial interfaces, intended for inter-chip connections. These types of I/O are very specific to a particular processor chip, and while they may require a lot of programming effort, they don't require much effort in the way of hardware design. However, interfacing an I/O device to a processor data bus is a significant process that is equivalent to the memory to processor interface design, and is subject to the same timing and loading analysis.

Direct CPU I/O Interfacing

The processors I/O pins may often be connected directly to simple devices, such as key switches and LEDs. In some cases an interface circuit may be required to convert the processor's I/O voltage and current levels to those appropriate for the I/O device. In order to understand which approach is appropriate, we'll investigate the capabilities of the processor's I/O pins, using the 8051 as the primary example.

Our objectives in this section are to understand how the I/O port circuitry is designed, how to interpret the relevant specifications, and the capabilities limitations of the circuits. The 8051 Port 1 I/O pins will be used to illustrate the unique characteristics of the quasi bi-directional circuits. The I/O port DC specifications and absolute maximum ratings will be compared to the requirements for driving a simple LED circuit. In addition, the I/O voltage specifications will be explained and we'll examine related protective circuits.

The characteristics of an external device must be considered in both the hardware and software design. For instance, mechanical switches used for manual input to microcontroller-based designs are prone to contact bounce, which causes the connection to open and close several times within a few milliseconds. The programmer must ignore these bounce conditions to prevent multiple key actions.

Port I/O for the 8051 Family

The I/O ports are mapped into the SFR (*special function register*) address space of the 8051, using direct access to the upper half of the internal data memory, addresses 80 through FFh (h = hexadecimal). In this example, we will use Port 1 on the basic 8051 device, which is the easiest port to describe since it has no alternate functions. For example, Port 1 is mapped to internal location 90h. This port can be used for general purpose I/O. Port 1 also appears in the bit addressable space as locations 90h to 97h. Port 1's LSB (*least significant bit*) is available at address 90h, and the MSB (*most significant bit*) is at address 97h in the bit-addressable space.

Port 1 on the standard 8051 family parts can sink a few milliamperes, however it can only source only 10 to100 microamperes. The entire port can be reset to zero by moving the value zero to location 90h by executing the instruction: MOV 90h,#0. The MSB (P1.7) could be set to logic one by setting bit number 97h executing the following instruction: SETB 97h. Bit P1.7 can be cleared to logic zero by executing the instruction CLR 97h. Likewise, a single input bit can be tested using a conditional jump instruction (such as JB 90h,address) that will jump to the address only if the LSB of Port 1 (P1.0) is high when the instruction is executed. You can easily observe this operation by using a logic probe or meter connected to pin 1 of the processor chip, which is the LSB of Port 1 (P1.0). The I/O pins will be in the logic one state after reset, but executing the CLR 90h instruction will clear P1.0. I/O pins can also be input directly to another bit, such as the carry bit, which is very useful when sending and receiving information by a serial bit sequence. This is a useful way to transfer data and addresses between the processor and serial I/O and memory devices.

For example, to output eight bits to Port 1 the following instructions can be used:

```
MOV    90h, A   ; Accumulator is output to port 1

MOV    P1, A    ; same as above, using the symbolic name for port 1

MOV    P1,0ffh  ; Output FF hex (all ones) to port 1
```

It is also possible to output a single bit, as shown below:

```
CLR    P1.0     ; The LSB of Port 1 is cleared (made equal to 0, ~0 Volts)

SETB   P1.0     ; The LSB of Port 1 is set (made equal to 1, ~5 Volts)
```

Likewise, eight bits can be input into the accumulator, using:

```
MOV    A, P1    ; Acc<=port 1
```

Single bit input can be accomplished from Port one bit 1 to the carry bit:

```
MOV    C, P1.1 ; Carry bit is loaded with the current state of P1.1
```

An input bit can also be used to control program flow:

```
JB     P1.0, address ; Jump to address if bit P1.0 is 1, otherwise continue
```

Monitor commands can also be used to access the I/O pins on the SDK:

```
#P1    allows direct R/W of port 1

#SB 92 allows observing and set/clr of P1.2 bit
```

Port 1 can be accessed one bit at a time in the bit addressable address space from 90h to 97h, which correspond to each of the eight bits of port 1. The MSB (P1.7) can be accessed at bit address location 97h. The entire port can be reset to zero by moving the value zero to location 90h executing the instruction:

```
MOV 90h,#0
```

The MSB (P1.7) could be set to logic one by setting bit number 97h executing the following instruction:

```
SETB 97h
```

Bit P1.7 can be cleared to logic zero by executing the instruction CLR 97h. Likewise, a single input bit can be tested using a conditional jump instruction, such as: JB 90h,address which will jump to the address only if the LSB of Port 1 (P1.0) is high when the instruction is executed. You can easily observe this operation by using a logic probe or meter connected to pin 1 of the processor chip, which is the LSB of Port 1 (P1.0). Normally the pin will be in the logic one state after reset, but executing the CLR 90h instruction will clear P1.0. I/O pins can also be input directly to another bit, such as the carry bit, which is very useful when sending and receiving information by a serial bit sequence. This is exactly how the data and addresses are sent and received between the processor and serial I/O and memory devices.

It's important to recognize that **some instructions modify the output latch, rather than the input pin.** This applies to instructions that read-modify-write the output pins, such as ANDing the port with a constant value to mask certain bits. This is necessary because the I/O pins can serve as input or output. Pins which are to be used as inputs must be written with a logic one output first, so that an external device such as a switch to ground, can pull the line low. If the pins were used directly, then a pin that was being used as an input but just happened to be low at the time that the logical AND operation was carried out, would become stuck low! By performing the logical AND with the output register instead, the state of the input pin will not be affected.

The internal circuits for the I/O pin are shown in simplified form in the figure. The 8051 uses a modified open-drain output structure, which allows it to operate as either input or output, or even both at the same time. It consists of a constant current pull-up (current source), an N-channel MOSFET switch as a pull-down device (FET sinks current). The FET is an active switch, so it can sink more current. That is why the 8051's sink current is large compared to source.

The simplified I/O port circuit diagram in Figure 8-1 shows a pull-up resistor providing a weak current source, and a FET pull-down capable of sinking more current. The input pin can also be read

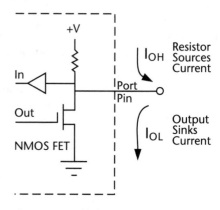

Figure 8-1: Simplified I/O port circuit.

by the input buffer. This allows the pin to be used as either input or output. When using the pin as an input, the FET must be turned off by writing a one to the output pin. Then an external device, such as a switch connected between the pin and ground, will pull the input low when the switch is closed. When using one of these pins as an input, an external pull-up is usually not required, as the pin is pulled up internally.

While the simplified representation approximates the behavior of the circuit, in order to thoroughly understand how it behaves, we must go deeper. The diagram in Figure 8-2 shows a somewhat more accurate version of the circuit, which is referred to as a "quasi-bi-directional" circuit. The pull-up is actually a current source, which can source one of two currents. When the output is static in the high state, the current source provides about 50 microamperes of current to an external load.

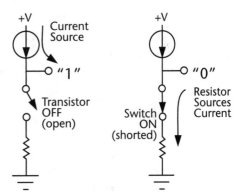

Figure 8-2: Quasi-bi-directional pin.

When the output pin transitions from one to zero, the FET switches on, sinking the source current and the current from any output load to ground. The switch is not perfect, and has some resistance, which causes the output voltage to rise somewhat above ground. If the current source is a resistor, then the low-to-high output voltage transition would be very slow, due to the R-C time constant formed by the resistor and the load capacitance. Even with a small constant current source, the output voltage will ramp up slowly. The current source in the 8051 behaves differently on a zero-to-one transition. When the output pin transitions from zero to one, the current source provides a much higher current for a very short time, pulling the output voltage up quickly. Then the current source reverts to its lower value. This unique feature of the output addresses the slow rise time problem by lowering the time constant during the zero-to-one transition, without requiring an external input device to sink more than 50 microamperes. A secondary benefit is that the pin circuitry does not have to be explicitly programmed as an input or output, as is the case with all other microcontroller families. This also means that the pin can be used alternately for input and output, like an open-collector or open-drain bus without concern for bus contention. This is useful for things like shared request lines and multiprocessor communication. The disadvantage to this type of I/O circuit is that it cannot source much current. The sink current is greater than the source current, but still less than other microcontrollers.

Output Current Limitations

The output low (sink) current for the 80C32 is limited to approximately15 milliamperes maximum. That is an *absolute maximum* specification value, meaning that output current in excess of this value can damage the device. Shorting a low output to the power supply would damage the device. In addition, the total sink current for an 8-bit port is limited to approximately 26 milliamperes. So if all the outputs of a port are low at the same time, they can only sink a little more than 3 milliamperes each.

On the other hand, the current source will not supply any more than about 50 microamperes under static conditions, so it cannot be destroyed by shorting an output to ground. The 80C32 current source also has an additional feature that improves input noise immunity. The current that must be sunk by an external device trying to pull the 80C32 pin low increases as it approaches ground during a one-to-zero transition. That means that weak low going noise pulses are less likely to cause an error.

Let's examine a simple case, that of driving a LED which needs around 10 milliamperes to be clearly visible. In this case, we connect the LED and a resistor to limit the current between the power supply and the processor pin as shown in Figure 8-3.

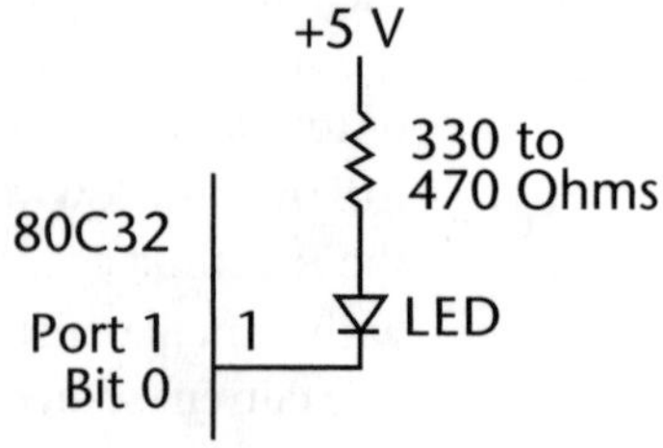

Figure 8-3: Driving a LED directly from a port pin.

The LED will be off as long as the output pin is high. When the output pin goes low, the output will sink current and the LED will turn on. LEDs have a relatively constant voltage (1.5 to 2 volts typical) across them when they are operating. If the LED has 2 volts across it, then the resistor has the remaining 3 volts across it, then the current in the resistor and LED is 3 volts/330 ohms, or about 9 milliamperes. This will be enough current to light the LED, but it won't be very bright. Also, the processor would only be capable of lighting a couple of LEDs. When more output current is required, other circuits can be used.

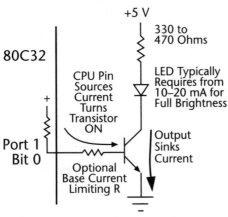

Figure 8-4: NPN transistor for greater load current.

Figure 8-4 shows how an NPN transistor can be used to amplify the current from the processor's output. The processor's output source current and transistor gain limit the potential load current. A special type of transistor, called a *Darlington transistor*, has a very high current gain, on the order of thousands. The CPU's output high current is multiplied by the transistor's gain, allowing much more current to flow in the load.

In this case, the 50 microampere source current is multiplied by the transistor gain, allowing more current to flow in the transistor collector, and hence the resistor and LED. When the output pin is high, the LED is on. For 8051 family parts, a current limiting resistor in series with the transistor base is not required, since the current source limits the base current. Other processor outputs will usually require the base resistor to limit the base current. The low source current and transistor gain is a limiting factor in this case, along with the higher saturation voltage on the collector-emitter output of the Darlington transistor compared to a regular transistor. Note that the output voltage switched by the transistor is separate from the processor supply, so this circuit can also be used to switch much higher voltages, limited only by the transistor's maximum collector voltage specification. Yet another approach, using a PNP transistor may be a better solution for high current loads.

This approach is shown in Figure 8-5. Using a PNP transistor so that the processor's output greater output low sink current to turn on the transistor, allows a standard transistor to be used in place of a Darlington device. It also allows the output switch to control a grounded load, which the previous versions could not. For an output low current of 1.6 milliamperes (one standard TTL load) and a

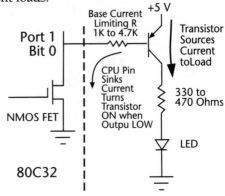

Figure 8-5: PNP transistor output driver.

modest transistor gain of 50, the transistor will be switched on with very little voltage across the transistor. Note that the LED will be on when the I/O pin is low. When the processor is reset, all the output pins are set high. This is good for loads that must star out without power when the device is first powered up.

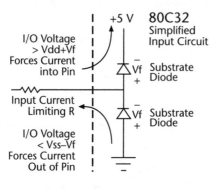

Because of the way the transistor is connected, this configuration does not allow the load to be connected to a supply voltage higher than that of the processor's. By combining the NPN and PNP transistor circuits, it is possible to switch higher voltages. Higher voltages can cause problems on the input pins if not properly protected. The reasons for this are illustrated in Figure 8-6.

Figure 8-6: I/O pin voltage limits.

Looking at the absolute maximum ratings for a chip, you will observe that most device inputs must be kept within a diode's forward voltage drop of the power supply and ground. When turned on, a silicon diode has about a 0.6 to 0.7 volt drop across it. There are parasitic diodes from the input pins to the power and ground signals, which are used to isolate the various internal circuits on the chip from one another on the chip's substrate. The substrate is the foundation upon which all the transistors and other components are laid, and is usually also the signal ground. The diodes can be turned on if the input goes above the power supply or below ground, causing large currents to flow in the chip. Even worse, these currents can cause a CMOS chip to "latch up," damaging or destroying the chip. This occurs because CMOS chips have four layers, equivalent to a silicon-controlled rectifier (SCR), which shorts its outputs as long as power is applied, once it has been triggered. The net effect is that the CMOS chip will become a short between the power supply and ground, causing large currents to flow, quickly heating up and even burning out the entire chip. Generally this will occur in such a way as to burn out the most expensive chip on the board, thereby protecting the 10¢ power supply fuse from blowing out!

Voltages that exceed the chip's allowable limits can be generated by overshoot on the signals due to unterminated transmission lines, electrostatic discharge (ESD) effects, or power transients. It can also be caused when an unpowered device's inputs are driven by a separately powered device. When power is applied to the previously unpowered device, having the inputs at a higher level than the supply voltage can cause latch-up. By using a resistor in series with the input, as shown in the previous figure, it is possible to limit the current in these conditions to a level which will not cause latch-up to occur.

The 80C32 parameters are different than other members of the 8051 family. The Atmel 89C2051, a low cost 20-pin version, has greater output drive capability than the 80C32. Depending upon which port is used and how it is configured, the output capabilities can also vary, even on the same device.

Processors other than the 8051 family of devices frequently have different characteristics, including: standard tri-state outputs with higher drive capacity and data direction control registers, and much higher output source and sink currents. For example, the Microchip PIC family of processors has devices that are capable of sinking *and sourcing* up to 25 milliamperes per pin. Note that the price for the higher drive capability is the requirement to write to the data direction register for bi-directional I/O functions, and the potential for bus contention problems. Higher output drive on any microcontroller can be accomplished using external power control devices, designed for driving motors, solenoids, valves, and other larger loads. Some of these devices have additional features, such as current limiting, over temperature shutdown, and so forth. Some also have limited logic built in, and are often referred to as "smart power" devices.

There are several common types of I/O device which can be directly connected to the processor, including simple switches, keypads, LEDs, and LCDs. Input devices can be divided into three categories: simple switches, multiplexed keyboards, and intelligent keyboards as used on the desktop PC. The displays can also be divided into three groups: simple on/off indicators, multiplexed LED or LCD displays, and intelligent display modules. People can also be classified into three groups: those who divide things into groups, those who do not, and those who have no opinion.

Simple Input/Output Devices

The switch is probably the simplest of all input devices, and one of the most useful. Hardware interfacing is quite simple, and for CPUs that have internal pull-ups like the 8051, all that need be done is connect the switch between the pin and ground. As can be seen from Figure 8-7, the input will be a logic one when the switch is open, and logic zero when the switch is closed. Unfortunately switch contacts bounce when they are closed and sometimes when they are broken. This causes the output to oscillate briefly between one and zero until the contacts stop bouncing, usually after several milliseconds or more. As a result, the program reading the switch state must "de-bounce" the switch operation, meaning that the switch transitions must be ignored for some time after the first transition between off and on.

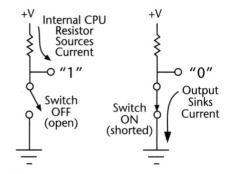

Figure 8-7: Simple switch used as input.

Matrix Keyboard Input

The next step up in input complexity is the matrix keypad or keyboard. These switch arrays are usually organized into a number of rows and columns, like the 4-by-3 array of 12 buttons on a telephone. These matrix-connected devices can be multiplexed to reduce the number of I/O lines required to sense the keys. If a 4-row-by-4-column keypad were implemented using separate inputs, one per switch, a total of 16 input pins would be required. Since I/O pins are almost always at a premium, this is not the best approach.

By arranging the switch contacts to short the row and column lines corresponding to their position in the matrix, the number of lines can be reduced. By selecting one column at a time and looking for activity on any of the row inputs, the program can determine which key has been depressed. One row output can be driven low at a time, and the column input bits are read to see if any of them are low. A low column input indicates that the switch belonging to the corresponding row and column is closed. Multiplexing allows the rows and columns to be scanned for activity under software control. In the case of sixteen keys, only four columns and four rows would be required, or a total of eight I/O pins, compared to 16 for the simple one input per switch approach. For the processors like the 8051 with built-in pull-ups, the only thing that is required is the key switch matrix. A key switch matrix like this can be implemented very inexpensively by using a standard matrix keypad, or by attaching steel switch domes to a PC board with row and column contacts, encapsulated under an adhesive plastic sheet. An adhesive backed label can be printed using a standard printer and covered with another layer of clear plastic. The resulting keyboard will have a custom graphic legend and be relatively impervious to contamination. The cost of this type of "click dome membrane keypad" is also very low, whether you make it yourself or buy one from a manufacturer that specializes in these keyboards. Figure 8-8 shows the schematic of a multiplexed keyboard matrix.

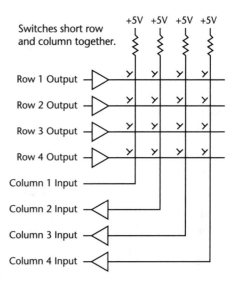

Figure 8-8: Matrix keypad multiplexing.

Even fewer I/O lines can be used if the rows are decoded using a 2-to-4 line decoder, and the columns are encoded using a 4-to-2 line priority encoder. This approach will require only four I/O pins. Using a 3-to-8 line decoder and an 8-to-3 line priority encoder, it is possible to scan 64 keys using only six I/O lines.

A multiplexed keyboard can also be scanned using a dedicated matrix keyboard IC, such as the 74HC922, which provides hardware controlled scanning automatically as well as a separate interrupt output. This device can be mapped into the external memory or I/O address space of a processor.

Matrix Display Devices

Simple output indicators, such as the simple LED indicators presented previously, can be very useful, but similar problems arise when using multiple LEDs each driven by a single I/O pin. Once again, the LEDs can be arranged in a matrix, and driven by multiplexing rows and columns of devices. As long as the LEDs are scanned quickly enough, at least 15 or 20 times per second, the LEDs appear to be on continuously. This works because of a perceptive characteristic of human vision, known as *persistence of vision* (POV). Many devices, including television and computer CRT displays depend on this characteristic of human vision. Many LED display alarm clocks use multiplexing, as do many other types of displays, such as most LCDs. In each case, the flicker of the display is normally not apparent to the observer. You can see the strobe-like effect by waving your fingers quickly in front of a multiplexed display.

An array of LEDs or seven segment numeric LED displays can be illuminated this way, using many fewer I/O pins than would be required by using one pin per LED, as shown in Figure 8-9.

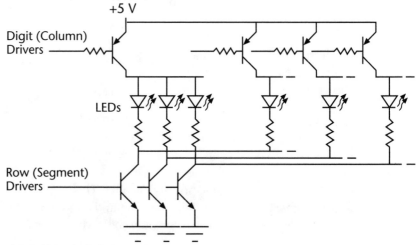

Figure 8-9: Multiplexed LED display.

The display is scanned, or refreshed, by activating the column, and then the row bits that correspond to the LEDs in that column which should be lit. The display is left on for a short period, then switched to the next column and row, and so on. As long as the display is refreshed frequently enough, there is no visible flicker.

Another type of display is the LCD. The simplest of these is just a glass panel with extremely thin metalized connections to the segments. These are rather complex to drive directly from most microcontrollers, but there are two ways that they can be connected without much effort. The simplest, but more expensive approach, is to use an intelligent LCD module complete with the drive electronics. Most of these devices use a standard controller, and can be driven using either a 4-bit bus or an 8-bit bus. Serial input devices are also available, which can be driven directly from a standard RS-232 serial port. They are available in text-only display versions, ranging in size from one row of 16 characters to four rows of 40 characters. Graphic display versions of these modules are also available, allowing flexible text and graphic display formats.

Another method of driving small glass displays directly is through special LCD display driver chips, which are designed to drive a relatively simple display (such as one containing simple 7-segment numeric digits, for example). These peripherals are available from several vendors, and the LCD display peripheral driver hardware is even incorporated in some microcontrollers.

Many other types of I/O can be added externally using the processor's bus interface. The 82C55 chip is a commonly used parallel interface with two 8-bit ports and two 4-bit ports which can be programmed as inputs or outputs. Connecting an 82C55 to the 8051 bus using memory mapping is an example of a program controlled I/O interface.

Program-Controlled I/O Bus Interfacing

In this form of I/O, the processor communicates with I/O devices in essentially the same way it communicates with memory. The program running in the CPU must check the availability of data and transfer it, one piece at a time. The processor puts an I/O address on the bus, indicates the type of transfer, either read or write (I/O read or I/O write cycle for processors with an I/O address space). The CPU uses activates its control lines, and then transfers the data to or from the selected I/O device. The 8051 does not have an external I/O space, so these devices must be mapped into the external data memory address space. Processors with a separate I/O address space, such as the x86 family, have input and output instructions that cause the CPU to generate the appropriate I/O read and I/O write instructions respectively. Processors with a single address space, such as the 68000 family, have no I/O instructions. They use memory mapped I/O, so both software and hardware treat the I/O addresses in the same way as memory.

An *I/O interface* connects the actual I/O device, such as an LED, a switch or a printer, to the CPU. The job of the designer is to design an interface that meets the requirements of both the I/O device and the bus. While memory devices only read or write data, I/O devices may perform other operations as well. A typical I/O interface has several addresses, usually referred to as *I/O ports* or *I/O registers*, for different types of information such as data, commands, and status. These registers are the "window" through which the programs must monitor, control, and communicate with the corresponding I/O device. Three types of information are typically exchanged through this window: commands from the CPU to control the I/O device, status of the I/O device to the CPU, and the actual data to be transferred. Many interfaces have I/O registers corresponding to these three types of information as follows:

- **Command Register.** This is sometimes referred to as the *control register*. This register is written by the CPU to control things such as the operating mode of the I/O device, direction of data transfer, enabling or disabling the use of parity, interrupts, and so on. Usually each bit or field of bits is used to control a specific function, but the commands may also be encoded in a way equivalent to that used for encoding information in the CPU instruction op codes. Several of these "control words" may be required to initiate I/O operations. Control words written to the command register would be instructions to the I/O interface on how to perform a specific type of transfer. In some cases the command register is "write-only," meaning that the information that is written into this register cannot be read back by the CPU.

- **Status Register.** This register indicates the state of the I/O device at the time the register is read. The bits in this register typically indicate things such as the availability of data to be input as from a keyboard, or output as to a printer. By reading the status register, the program running in the CPU can determine when to transfer data and the presence of errors, among other things. Typical status bits would be "input data ready," or "output data register full." Sometimes the status register is "read-only," meaning that the information in this register can only be controlled by the I/O interface and cannot be written to or modified by the CPU.

- **Data Register.** This register contains the actual data to be transferred to or from the I/O device. In some cases two separate registers and I/O addresses are used for input and output data, but in most cases they share the same address. Reading or writing information to this register will generally affect one or more status bits indicating the availability of data for the CPU or the I/O device. For example, when the I/O device has data ready for input, it would set the "input data ready" bit of the status register, and when the CPU reads the data register, the "input data ready" bit would be reset.

The process of testing a ready status bit is referred to as *polling* the device to see if it is ready for data transfer. Before any data can be transferred, the status register must be polled to determine if the device is ready. If the program is written to loop continuously waiting for the device to become ready, a lot of CPU time is wasted if the data is not available shortly after the polling begins. An example would be a keyboard, where keys are pressed at relatively slow and unpredictable rates. In order to minimize the time wasted in polling for these irregular data, interrupts are used. An interrupt is triggered by an event that is not synchronized to the main program and calls a special subroutine, referred to as an *interrupt service routine* (ISR that transfers the data. This "on-demand" processing is more efficient when data rates are relatively slow or unpredictable. At the other extreme however, when peak data transfer rates are high as they are in a disk drive, another technique that reduces the amount of work the CPU must do to transfer I/O data is used. The I/O interface transfers data directly between the I/O device and memory without CPU intervention using *direct memory access* (DMA).

Real-Time Processing

Some applications demand that the CPU respond to external events and process them in a finite amount of time. *Real-time processing* means that data are processed at the same rate that they occur. They are *event-driven* which means they are triggered by external events, such as the tick of a clock, completion of I/O, etc. Examples of real-time PC programs are the flight control program on the Space Shuttle, arcade games, speech processing software, and flight simulators. Examples of non-real-time PC programs would be word processors and accounting programs.

Direct Memory Access (DMA)

Direct memory access (DMA) requires that the I/O interface be active and semi-intelligent, since it must count the words and increment the memory address for each element transferred in addition to performing the actual transfer.

The transfer process involved with DMA is typically as follows:

1) The program writes into the I/O control register of the interface:

    a) The type of transfer (I or O).

    b) The number of bytes or "block size" to transfer.

c) The physical address in memory where the data will be transferred.

d) A start command is given to begin the transfer.

2) Data is transferred directly between memory and I/O devices under control of the I/O interface.

3) When the transfer is complete, the I/O interface sets a completion bit in the status register, and may also initiate an interrupt to the CPU.

Figure 8-10 compares program-controlled and DMA I/O.

Direct memory access is used for high speed I/O. The I/O device interface takes over the bus from the CPU and controls the transfer of data between memory and I/O directly, without any intervention by the CPU (as shown in Figure 8-10). Data is generally transferred in larger blocks, such as a disk file block.

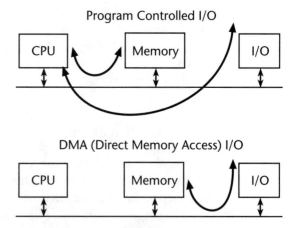

Figure 8-10: Program controlled versus DMA I/O.

Devices on a bus can talk with each other without talking with the CPU, except to tell it when done. DMA is good for disk and network transfers because the rates are much higher than the CPU can handle using program controlled I/O. There are two ways of doing DMA transfers: *single cycle DMA* and *burst DMA* modes.

Burst vs. Single Cycle DMA

In burst mode DMA, the DMA device gets control of the bus, transfers a whole block of data (a disk sector, for example), and then releases the memory back to the CPU. A single cycle DMA device gets the bus, transfers just one word of data, and releases the bus. *Arbitration* is the process of determining what device will have control of the memory bus.

Burst mode has low overhead and can handle the highest peak data rates, but the CPU can get locked out of memory for intervals that are as long as the longest block to be transferred. If the transfer is longer than the shortest interrupt interval, such as the real time clock tick interval, interrupts can get lost.

Cycle Stealing

In this mode, DMA transfers are completed during bus cycles that are not used by the CPU, so no arbitration needs to be done. Most modern, high performance processors utilize almost 100% of available memory bandwidth however, so there isn't much available for DMA. To save time, it is possible to perform arbitration and data transfer overlapping in time.

In general, burst mode DMA is more effective when relatively short time durations are needed to transfer the data block. Under those conditions, the bus is fully utilized for a short time interval. The DMA controller acquires access to the memory, transfers an entire block of data, and then releases the memory. An entire block of data is transferred in one short burst. The disadvantage is that a burst mode DMA device "hogs" the bus, thus preventing any other device from accessing memory during the burst. If the burst lasts too long, it may prevent the CPU from servicing certain time critical events, such as the real time clock interval (clock tick). In that case, the clock would run slower than it should because it would cause the CPU to miss some of the clock ticks. Therefore, burst mode DMA is most effective for data that is transferred at a high peak rate for short intervals. Typically, the data within a burst comes in too quickly to allow the arbitration handshaking required for the DMA controller to acquire and release the data between each data element. An example of this situation is the transmission or reception of data on a high-speed local area network interface. Small packets of data come across the network in high-speed (less than one microsecond per byte) bursts, with relatively low packet rates (milliseconds between packets).

For single cycle mode, the DMA controller acquires access to memory, transfers one word, and releases the memory. That allows other memory transfers to be interleaved with the DMA. That is why this mode is also referred to as "interleaved DMA." Single cycle DMA is better suited to transferring data over longer periods of time, where there is enough time to acquire and release the bus for every word transferred. In this case, the CPU and other devices can still access the memory, at a reduced bandwidth. As a result, the CPU may be a bit slower because it will sometimes have to wait for a DMA cycle to complete, but it is not entirely shut out when a DMA transfer is in progress. When a single cycle DMA transfer occurs, more time is used in acquiring and transferring control of the memory to and from the DMA controller since it happens so much more frequently. This "overhead" frequently reduces the overall available memory bandwidth, especially when it is performed sequentially with the data transfers. Some systems overlap the memory bus arbitration handshaking with the memory data transfers so that the arbitration does not slow down the data transfers.

Direct memory access is required when the CPU is too slow to transfer the data under program control. Because the CPU does not have to participate directly in the item-by-item transfer of data, DMA is also useful when there are other tasks that the CPU can perform. In those cases, DMA transfers may be used even though they are not strictly required by the data rate.

Elementary I/O Devices and Applications

Parallel ports are the simplest form of I/O, but there are many different types of electrical interfaces ranging from the simple open collector TTL outputs used on a PC printer port to high-speed peripheral interfaces such as the IEEE-488 and SCSI buses. Most embedded controller ICs have some pins that are configurable as parallel input or output. These interfaces are appropriate for simple I/O, such as key switch and display interfacing. They are also appropriate for controlling and monitoring high-level interfaces such as solid-state relays.

The parallel I/O ports available on the 8051 family and similar processors are fairly versatile, with special internal circuitry to allow a port bit to be configured individually as an input or output. Some microcontrollers also provide considerable current source and sink capability, however the 8051 family parts are usually fairly weak in that regard.

Serial ports, also referred to as asynchronous or synchronous communications (COM) interfaces, are commonly used to interconnect with devices, such as modems, which inherently transmit the data one bit at a time over a communication link such as a phone line. The RS-232 serial interface used in a PC's COM port is an asynchronous serial data stream. An asynchronous interface has no explicit clock signal to synchronize the transfer of data. The timing of bits is based on the absolute bit rate, and is synchronized on every character with a start bit. The serial to parallel conversion is performed by a UART (*universal asynchronous receiver-transmitter*). When transmitting data, the UART appends a start bit before the data, shifts out the data LSB first, and adds a stop bit after the data. Once the transmission is complete, the UART sets a status bit indicating that the data has been sent and that it is ready to begin transmission of another character.

When receiving data, the UART looks for and synchronizes to the leading edge of a start bit. Then, it delays for one and one-half of a bit period, so that it samples the LSB in the middle of the bit period. Then, the UART delays one bit period, samples the next to the LSB, etc. until all the bits have been shifted in. Once all the data has been received, it is loaded into a buffer register and a status bit is set to indicate that the receive buffer contains a character and may be read by the CPU. In order for this method to work, the two UARTS at each end of the communication must have bit rate clocks that are accurate enough to guarantee that the data will be sampled at the right time. This typically requires a sample clock that is 16 times the data rate, accurate to 1% to 2%.

Timers and *counters*, which are present in most microcontroller chips, allow generation of pulses and interrupts at regular intervals. They can also be used to count pulses and measure event timing. Some of the more sophisticated versions can measure frequency, pulse width, and relative pulse timing on inputs. Outputs can be defined to have a given repetition rate, pulse width, and even complex sequences of pulses in some cases. In most cases, one of the timers can be used to generate the necessary serial clocks required to operate a microcontroller's on-chip UART. In order to meet the approximately1% clock frequency accuracy for the 16x data rate clocks, the crystal frequency is often chosen to allow exact integer division of the crystal frequency resulting in an accurate, standard serial data rate. This is why 8051 family parts that use their internal counters and serial port to connect to standard 9600 bps and higher data rates use the crystal frequency 11.059 MHz rather than an even 12 MHz.

Analog to digital converters (ADCs) and digital to analog converters (DACs) are used to convert continuously variable real world parameters to digital form and back to analog. Examples include conversion of the output voltage of a temperature sensor into digital form for processing, and converting control values back into analog form to adjust the temperature. Most of the quantities of interest in the real world tend to be continuous and analog in nature, so these converters are critical for many applications.

Timing and Level Conversion Considerations

Depending upon the rate and load on the processor, peripherals can be interfaced using interrupt driven, program controlled, or DMA I/O. High speed devices will generally require DMA, while devices that generate small amounts of data at unpredictable times are better handled with interrupts and program controlled I/O.

Level Conversion

Many types of devices that need to be interfaced to the processor are not compatible with standard logic levels. For example, many serial interfaces comply with an interface standard, such as the EIA RS-232 specification, which defines the voltage level and pin out. RS-232 levels are nominally plus and minus 12 volts, instead of the 0 to 5 volt levels that most processors use. As a result, level shifting devices are needed to translate between the 0 to5 and +/-12 volt signals. Single ICs that provide the translation as well as generating the +/-12 volt supplies from a single +5 volt supply, are now available (Maxim MAX232 and others), making this much easier for embedded system designers.

Intermediate DC voltages can often be handled using simple open-collector outputs, or a separate transistor and pull up resistor to drive output voltages higher than the logic supply. Power switching FETs are also available that can handle relatively high currents and voltages, and can be driven directly by logic-level outputs.

Power Relays

High-level outputs, such as 110 volt AC loads, must be switched using solid state or magnetic relays. The magnetic relay windings are inductive coils that must be clamped using a diode to prevent large inductive transients from damaging the relay driver circuits. Solenoid valves and other devices are used to control external flow and have similar inductive characteristics. Solid-state relays are much easier to use, as they are isolated from the high voltage and provide a simple logic level interface. Optical isolation is also used to sense high voltage inputs and convert them to logic levels. There are even standardized modules (OPTO-22 and equivalents) available that can be interchanged with each other, resulting in very flexible configuration options.

Chapter Eight Problems

1. Using an 8031 Port 1 I/O bit, design an interface to an LED that requires 20 milliamperes of output current for full brightness.

2. A DMA device transfers blocks of data consisting of 256 bytes, and the bytes in the burst are spaced 10 microseconds apart. The real time clock tick interval is 1 millisecond. What kind of DMA should be used, burst mode or single cycle?

3. If an 8031 CPU executes one instruction per microsecond, estimate the maximum rate that data can be transferred to or from an I/O port, assuming that a status bit must be polled before transferring data.

4. Design a 4-row by 3-column telephone keypad matrix for connection to the 8051 Port 1 pins, to be polled using software scanning.

Other Interfaces and Bus Cycles

There are two kinds of interrupts software and hardware. *Software interrupts* are just another kind of subroutine call that can be used to access subroutines with entry points at fixed memory locations. Operating system services are often accessed using software interrupts, which are simply instructions that cause an interrupt subroutine to be called at whatever point in the program they are placed. These interrupts are synchronized with the program in that they always occur at the same place in the program. They are referred to as "synchronous events because their execution is solely dependent upon the sequence of execution of the program instructions.

Some processor manufacturers refer to "traps" or "exceptions," but these are synonymous with the term "interrupt" as used here, which may be either a hardware or software interrupt. Unless otherwise specified, however, the word "interrupt" is generally used to imply a hardware interrupt. *Hardware interrupts* are triggered by a physical event, such as the closure of a switch, that causes a specific subroutine to be called. They can be thought of as a sort of hardware initiated subroutine call. They can and do occur at any time in the program, depending on when the event occurs. These are referred to as "asynchronous events because they may occur during the execution of any part of the program. Interrupts allow the programs to respond to an event when it occurs. In a printing application, the printer may interrupt the processor to inform the program that it has printed all the data in its buffer and is ready for more. A serial interface might activate an interrupt to indicate that a character has been received and it is available to be processed. These kinds of applications are "event driven" because no action will take place until an event occurs. In the case of a typical embedded application, event driven programs are used when it is necessary to respond to an external event within a fixed time period. A system that has to respond to, and then process, an event in a fixed amount of time is referred to as a *real-time* system.

Interrupt Cycles

When a hardware interrupt request is enabled and activated, the CPU saves its current program counter and performs an interrupt cycle in place of the usual program fetch cycle. The interrupt cycle typically consists of the interrupt source identification and the transfer of the interrupt vector information. The interrupt vector is often a pointer to the place in memory where the address of the interrupt service routine is stored. The CPU will then fetch that address and perform what amounts to a subroutine call to that address. When the interrupt subroutine has completed processing of the event that caused the interrupt, the processor executes a "return from interrupt" instruction and goes back to the part of the main program that was executing before the interrupt occurred.

Software Interrupts

A software interrupt is a special subroutine call. It is *synchronous* meaning that it always occurs at the same time and place in the program that is interrupted. It is frequently used as a quick and simple way to do a subroutine call for accessing programs such as the operating system and I/O programs. A disk operating system used on the PC uses interrupt number 21 (hex) to invoke operating system functions such as reading a disk file, output data to the printer and so on. For the purposes of this chapter, the word "interrupt" by itself will be taken to mean a hardware interrupt.

In Chapter Six, we looked at the 8051 processor's program read, data read, and data write cycles. However, most processors also have other types of bus cycles, including special cycles for processing hardware interrupts.

Hardware Interrupts

A hardware interrupt can be thought of as hardware induced subroutine call. When an external event—such as the pressing of a key—occurs, an interrupt subroutine is called to store the key code for later use. This type of *event-*

driven subroutine call is asynchronous meaning that it can occur at any time and place in the program that is interrupted. Interrupt *latency* is the term used to describe the amount of time from when an event occurs (such as the pressing of a key) until the interrupt subroutine begins execution.

Among the factors that determine latency are:

- Hardware, which determines the time the CPU requires to process the request and acknowledge sequences (fixed hardware time).

- The time required to get a vector and load it into the processor ("vectored" interrupts will be discussed later in this chapter).

- Sequences of code with interrupts disabled add directly to latency.

- Higher priority interrupts overriding the current interrupt (the time a high priority takes to execute adds directly to the latency of lower priority interrupts). This is reduced by keeping ISRs as short as possible.

Figure 9-1 shows a common hardware interrupt situation—handling a pressed key on the PC's keyboard. Figure 9-2 shows the timing sequence for handling the interrupt service routine (ISR) required to process the key press event.

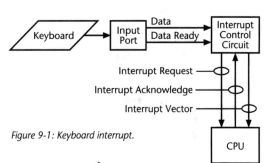

Figure 9-1: Keyboard interrupt.

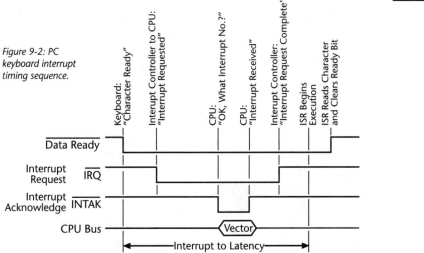

Figure 9-2: PC keyboard interrupt timing sequence.

Interrupt Driven Program Elements

When an interrupt is processed, here is a detailed sequence of typical elements involved:

1) **Initialization** executed once)
   a. Disable interrupts (*always do this first!*).
   b. Clear buffers/ pointers/ flags.
   c. Store address of ISR(s) in vector table.
   d. Initialize interrupt hardware.
   e. Clear any interrupt requests.
   f. Enable interrupts and enter MAIN routine.

2) **Main routine** (executed many times, when no interrupts are pending)
   a. Performs processes that are not time critical, such as diagnostics.
   b. Access any resource that is NOT re-entrant.
   c. Wait for interrupts to occur.

3) **Interrupt service routine (ISR)** executed once per interrupt)
   a. Save processor state: registers, flags, interrupt level, etc.
   b. Process the event (what we really wanted to do in the first place).
   c. Restore processor state: registers, flags, interrupt level, etc.
   d. Enable interrupts (may be located at different points in the ISR, depending on requirements).
   e. Tell interrupt controller we are finished processing this interrupt return from interrupt.

Re-entrant code or a re-entrant routine is code that can be interrupted at any point when partially complete, then called by another process, and later return to the point where it was interrupted to complete the original function without any errors. Non-re-entrant code, however, cannot be interrupted and then called again without problems. An example of a program that is not re-entrant is one that uses a fixed memory address to store a temporary result. If the program is interrupted while the temporary variable is in use and then the routine is called again, the value in the temporary variable would be changed. When execution returns to the point where it was interrupted, the temporary variable will have the wrong value. In order to be re-entrant, a program must keep a separate copy of all internal variables for each invocation. Re-entrant code is required for any subroutines that must be available to more than one interrupt driven task.

Interrupts can be processed between execution of instructions by the CPU any time they are enabled. Most CPUs check for the presence of an interrupt request at the end of every instruction. If interrupts are enabled, the processor saves the contents of the program counter (PC) on the stack, and loads the PC with the address of the ISR. Some CPUs allow certain instructions to be interrupted when they take a long time to process, such as a block move instruction.

Critical Code Segments

Let's suppose there are two processes that both require occasional use of the printer. In a system that allows a task to be interrupted at any time by another task, simple binary flags will not be reliable. In the example below, two tasks are contending for access to the printer. The flag indicates whether the printer is in use, and is set equal to one to signal other tasks to wait until the printer is available. The premise is that each process will wait until the printer is free before attempting to print. Unfortunately, in an interrupt driven system, that will not always work. The example below shows when it can fail.

```
Process A:                          | Process B:
1 ACC := Flag                       | ACC := Flag
                                    |
2 Test ACC=0? if not,               | Test ACC=0? if not,
     go to start ^                  |      go to start ^
                                    |
3 if ACC=0:                         | if ACC=0:
                                    |
4 Flag := 1 "printer in use"        | Flag := 1 "printer in use"
     Access printer                 |      Access printer
Flag := 0 "printer                  | Flag := 0 "printer
not in use"                         | not in use"
                                    |
Printer Flag = 0 not in use         |
             = 1 in use             |
```

Notice that if process A is executing instructions 1 to 4 and is interrupted by process B, then there will be two copies of the flag, one in process A's accumulator and another in process B's accumulator. As a result, both processes will test the flag in their local accumulator, set the flag, and proceed to use the printer.

The output on the printer from the two processes would be intermixed, even though each process appears to have exclusive access, from the data available to each process. The problem occurs because there are two copies of the flag. The sequence of instructions 1 through 4 cannot be interrupted without the potential of improper operation. Such a sequence is referred to as a *critical code segment* that cannot be interrupted without risk of producing incorrect actions.

Semaphores

One way to fix the critical code segment problem in the preceding paragraph problem would be to disable interrupts before instruction 1, and re-enable them after instruction 4. While this will solve the problem, this solution adds to interrupt latency. A more efficient solution is the use of a *semaphore* instead of a simple binary flag. A semaphore is a multiple state variable that can be tested and set in one operation (the test and set operation cannot be interrupted). Here is an example of using a semaphore:

```
Process A                        Process B

Start:                           Start:
INC flag;                        INC flag;
look for FF => 0 change          look for FF => 0 change
        if result non-zero               if result non-zero
              then DEC flag                    then DEC flag
              go to Start                      go to Start
        else                             else
        if result = 0 then               if result = 0 then
        Use Printer                      Use Printer
        ...                              ...
Use Printer:                     Use Printer:
      (access the printer)             (access the printer)
      DEC   flag                       DEC   flag

printer semaphore: >= 0 printer in use
                   = FF hex, printer not in use
```

Note that the INC instruction has the ability to test and set the semaphore in one instruction. The semaphore is incremented and the status flags are set in the same instruction. Since an interrupt can only occur between instructions, there is only one instance when the semaphore variable makes the FF to zero transition. If other processes increment the semaphore they will increment from zero to one or more. The first process that increments the variable from FF hex to zero gets exclusive access to the printer. This is guaranteed because the test and set operation is an indivisible operation, which is the key characteristic of the protection mechanism of a semaphore. It is important to note that increments and decrements must be paired. The semaphore is more powerful than a flag because the processes can all share the printer resource under this scheme. Only the first process using a resource locks out all others. The first process seeing the FF to 0 transition gets the resource.

The 8051 only has one instruction that performs the necessary indivisible test and set operation, the "decrement and jump if not zero" or DJNZ. Most processors have instructions that can be used for the semaphore test and set operation.

Interrupt Processing Options

There are a number of variations in the way interrupts can be handled by the processor. These variations include how multiple interrupts are handled, if they can be turned off, and how they are triggered. Some processors allow multiple (nested) interrupts, meaning the CPU can handle multiple interrupts simultaneously. In other words, interrupts can interrupt interrupts. When multiple interrupts are sent to the CPU, some method must be used to determine which is handled first. Here are the most common prioritization schemes currently in use.

- **Fixed (static) multi-level priority.** This uses a priority encoder to assign priorities, with the highest priority interrupt processed first. This is the most common method of assigning priorities to interrupts.

- **Variable (dynamic) multi-level priority.** One problem with fixed priority is that one type of event can "dominate" the CPU to the exclusion of other events. The solution is to rotate priority each time an event occurs. This ensures that no interrupt gets "locked out" and all interrupts will eventually be processed. This scheme is good for multi-user systems because eventually everyone gets priority.

- **Equal single-level priority.** If an interrupt occurs with an interrupt, the new interrupt gains control of the processor.

Some types of interrupts can be turned on or off under program control. *Maskable* interrupts are those that can be enabled and disabled by the CPU. These are used for non-catastrophic events, such as a key being pressed. In contrast, *non-maskable* interrupts (NMI) cannot be enabled for disabled by the CPU. These are reserved for catastrophic events such as a power failure or parity error. Non-maskable interrupts are usually edge triggered (see next section) because we want to "remember" the event before it goes away.

Level and Edge Triggered Interrupts

An interrupt can be *level* or *edge* triggered. A level interrupt depends on the logic value, or level, when the interrupt signal is sampled by the CPU at the end of an instruction execution cycle. In contrast, an edge triggered interrupt occurs when a change, or edge transition, occurs in the sampled interrupt signal.

In level triggered interrupts, the interrupt request input signal is sampled by the CPU at the end of each instruction execution, as shown in Figure 9-3.

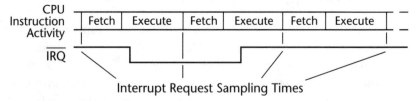

Figure 9-3: CPU sampling of level sensitive interrupt.

In this type of interrupt the IRQ line is sampled by the CPU, so there is a potential problem if the IRQ line goes active and inactive between samples. If the request goes away before it is sampled, the CPU will miss the interrupt. Also, if the interrupt request is still active when the processor has completed processing of the interrupt, it will be called and executed again.

The timing diagram of an edge triggered interrupt is shown in Figure 9-4. When there is an edge on an edge sensitive IRQ, it is latched inside the CPU until it is processed. Figure 9-4 shows an interrupt that is sensitive to falling edges.

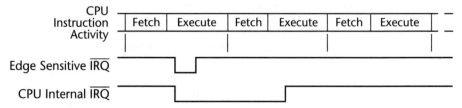

Figure 9-4: Edge sensitive interrupt.

It is possible to do the same latching with an external circuit to make a level sensitive interrupt into an edge triggered interrupt by using a flip/flop to latch the request as shown in Figure 9-5. When IRQ goes high, Q goes high until Clear pulses high, Q goes down.

When IRQ goes high, Q goes high until Clear pulses high, then Q goes down.

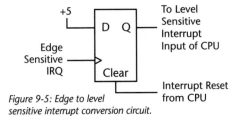

Figure 9-5: Edge to level sensitive interrupt conversion circuit.

As a general rule, use edge triggering when the interrupt pulses are very long or very short. Figure 9-6 shows a situation where the request pulses are very long, such as the 60 Hertz square wave that is often used for clock functions. A level sensitive interrupt input would generate multiple interrupts per 60 Hertz cycle. By using an edge sensitive input, there is only one interrupt since there is only one falling edge per cycle. Figure 9-7 shows the opposite situation: very short interrupt pulses. When the pulses are very short, the CPU could miss interrupts as shown below. An edge sensitive input will latch the interrupt until it can be processed.

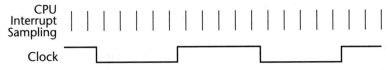

Figure 9-6: Long interrupt request cycles require edge sensitive input.

Figure 9-7: Short interrupt request pulses require edge sensitive input.

However, there are conditions where level triggering is preferable. When interrupt signals overlap, interrupts may be missed if an edge sensitive interrupt were to be used, as shown in Figure 9-8. This problem occurs on a machine where multiple interrupts are combined on one request line, as shown

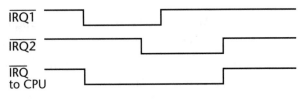

Figure 9-8: Overlapped requests require level sensitive input.

in Figure 9-9. This is typical of a microcomputer bus with shared interrupt request signals on the bus, and for devices that are capable of generating multiple interrupts simultaneously. This is often implemented by connecting multiple open-drain or open-collector, active low requests to the interrupt request line with a pull-up resistor. This allows multiple devices to use the same /IRQ line.

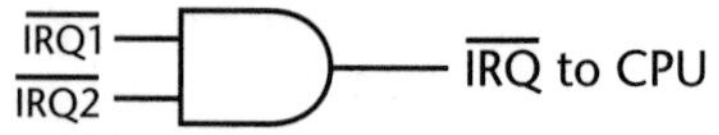

Figure 9-9: Multiple interrupts on a common bus.

An edge triggered system would sense only one edge, and thus it may miss IRQ2 whereas a level sensitive system will respond to both. An example of this condition in the 8051 CPU is the serial I/O port interrupt. The "receive buffer full" and the "transmit buffer empty" signals are combined as shown above to a common level-sensitive internal interrupt request. If the receive buffer happened to be filled and the transmit buffer emptied at the same time, there would only be one edge, due to the overlapping requests. Thus, a level sensitive input is required to guarantee that both interrupt will be serviced.

Vectored Interrupts

In a vectored interrupt system, the interrupt request is accompanied by an identifier, referred to as a *vector* or *interrupt vector* number that defines the source of the interrupt. The vector is a pointer that is used as an index into a table known as the *interrupt vector table*. This table contains the addresses of the ISRs that are to be executed when the corresponding interrupts are processed. The 8051 CPU architecture does have separate interrupt vectors for different interrupts, but it does not have an interrupt vector table. Instead, each interrupt is assigned a separate absolute memory address that will generally contain a jump to the actual ISR to be executed.

In other processors with interrupt vector tables, when a vectored interrupt is processed, the CPU goes through the following sequence of events to begin execution of the ISR:

1. After acknowledging the interrupt, the CPU receives the vector number.
2. The CPU converts the vector into a memory address in the vector table.
3. The ISR address is fetched from the vector table and placed in the program counter.

For example, when an external event occurs, the interrupting device activates the IRQ input to the interrupt controller that then requests an interrupt cycle from the CPU. When the CPU acknowledges the interrupt, the interrupt controller passes the vector number to the CPU. The CPU converts the vector number to a memory address. This address points to the place in memory, which in turn contains the address of ISR.

Non-Vectored Interrupts

For systems with non-vectored interrupts, there is only one interrupt service routine entry point, and the ISR code must determine what caused the interrupt if there are multiple interrupt sources in the system. When an interrupt occurs a call to a fixed location is executed, and that begins execution of the ISR. It is possible to have multiple interrupts pointing to the same ISR. The first act of such an ISR is to determine which interrupt occurred and branch to the appropriate handler. Serial I/O ports frequently have one vector for transmit and receive interrupts.

A typical microcontroller serial I/O port consists of a serial-in/parallel-out shift register for receiving serial input data, and a parallel-in/serial-out shift register for transmitting serial data, as shown in figure 9-10.

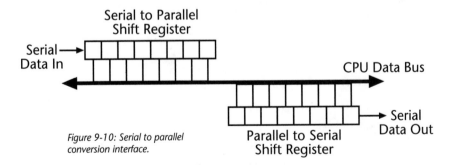

Figure 9-10: Serial to parallel conversion interface.

When the last bit of serial data shifts into the receive register, the receive interrupt bit is set (the RI SFR bit in the 8051) to indicate that the receiver buffer is full and ready to be read by the CPU. Likewise, the transmit interrupt bit is set (the TI SFR bit in the 8051) to when transmit buffer is empty and ready to accept more data from the CPU.

When multiple simultaneous interrupts occur, the processor must have some way of choosing which interrupt should be processed first. There are two common techniques for resolving the priority of simultaneous interrupts: serial and parallel.

Serial Interrupt Prioritization

When an interrupt occurs, the interrupting device lowers IEO and waits until IEI is high. Each device below it in line lowers its IEO. The device then performs an interrupt cycle. When the ISR is complete an end of interrupt occurs, the interrupting device raises its IEO line, which propagates down the line. This is usually referred to as a *daisy chain* interrupt priority system. At any given time, the highest priority device in the chain will be serviced first. Figure 9-11 illustrates this process.

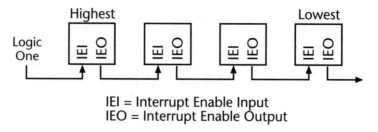

IEI = Interrupt Enable Input
IEO = Interrupt Enable Output

Figure 9-11: Serial "daisy chain" interrupt prioritization.

Parallel Interrupt Prioritization

A parallel priority encoder can also be used to prioritize multiple simultaneous interrupt requests. The priority encoder encodes the highest priority active input as a binary value, and that value is used as part of the interrupt vector number. The interrupts could be prioritized using an encoder that is equivalent to a 74x148 style 8:3 line priority encoder.

In most machines, the CPU checks for interrupt requests just after execution of each instruction. When an interrupt is enabled and occurs, the CPU will:

1. Save the PC (program counter) on the stack.
2. Acknowledge the interrupt request and get the vector from interrupt source.

3. Use the vector as an address or as a pointer into the interrupt vector table to fetch the address of the ISR from the vector table.

4. Load the address of ISR into the program counter.

5. CPU executes the ISR until return from interrupt execution at end of ISR.

6. Pop address off stack into program counter.

7. Continue execution where interrupt occurred.

The purpose of the interrupt processing sequence is to allow the processor to temporarily stop an executing program when an external event occurs, call the appropriate interrupt service routine to process the event, and then return to the interrupted program where it left off.

Interrupts provide a very efficient means for the processing of events that occur at unpredictable times with a minimum of delay. This is particularly important when there are a number of things that the processor must handle concurrently. Whole operating systems, usually referred to as *real-time operating systems* (RTOS), are designed to allow an application programmer to design multiple programs that can run concurrently on a single CPU almost as if they were running on separate processors.

Other Useful Stuff

This chapter surveys practical design issues that must be considered in an embedded design. Some of these topics are covered in more detail by the references in Appendix B.

Construction Methods

Embedded controllers can be constructed using any one of several techniques, but the most common method is a *printed circuit board* (PCB). The PCB is constructed of insulating material, such as epoxy impregnated glass cloth, laminated with a thin sheet of copper. Multiple layers of copper and insulating material can be laminated into a multi-layer PCB. By drilling and plating holes in the material, it is possible to interconnect the layers and provide mounting locations for through-hole components.

In designing the layout, or interconnecting pattern of the PCB, there are many conflicting requirements that must be addressed to make a reliable, cost-effective and producible device. For low speed circuits, the parasitic effects can be ignored and are often assumed to be ideal connections. Unfortunately, real circuits are not ideal, and the wires and insulating material have an effect on the circuit, especially for signals with fast signal rise/fall times. The traces, or wires, on the PCB have stray resistance, capacitance, and inductance. At high speeds, these stray effects delay and distort the signals. Special care must be taken when designing a PC board to avoid problems with transmission line effects, noise, and unwanted electromagnetic emissions.

Power and Ground Planes

When possible, it is a good idea to use two layers of a four or more layer PCB dedicated to the Vcc and ground signals. These are referred to as *power* and *ground planes*. One advantage is that there is a beneficial high frequency parasitic power supply decoupling capacitance, which reduces the power supply noise to the ICs. Power planes also reduce the undesirable emission of electromagnetic radiation that can cause interference, and reduce the circuit's susceptibility to externally induced noise. The power planes tend to act as a shield to reduce the susceptibility to external noise and radiation of noise from the system.

Ground Problems

While the concept of an ideal circuit ground may seem relatively simple, a great many system problems can be directly traced to ground problems in actual applications. At the least, this can cause undesirable noise or erroneous operation; at the worst, it can result in safety problems, including possibly even death by electrocution. Lest you dismiss the importance of this too quickly, the author has narrowly missed electrocution while testing a device in which the grounding was improperly implemented!

These problems are most often caused by one of the following problems:

- Excessive inductance or resistance in the ground circuit, resulting in "ground loops."

- Lack of, or insufficient isolation between, the different grounds in a system: earth, safety, digital and analog grounds.

- Non-ideal grounding paths, resulting in the currents flowing in one circuit inducing a voltage in another circuit.

The solutions to these problems vary, depending upon the type of problem, and the frequency range in which they occur. Usually they can be simplified to reducing the currents flowing in common impedances of circuits which need to remain isolated using a single point ground, and the prudent application of shields and insulation to prevent unwanted parasitic signal coupling.

Electromagnetic Compatibility

Electromagnetic compatibility (EMC) issues have become much more significant now that there are a large number of electronic devices which unintentionally radiate electromagnetic energy in the same frequency ranges used for communication, navigation, and instrumentation. Regulatory agencies—such as the Federal Communications Commission (FCC) in the United States, the Department of Communications (DOC) in Canada, and similar organizations in Europe—have defined limits to the amount of energy such electronic devices are allowed to emit at various frequencies. Even more stringent requirements are placed on life critical equipment, such as aircraft navigation and life support equipment, because of the sensitive nature of the applications. Among other things, these devices are required to provide a minimum level of immunity to externally induced noise (radiated and conducted susceptibility).

In solving an EMC problem, the first step is to identify the source of the noise, the path to the problem area, and the destination where the problem manifests itself. Once these three characteristics of an EMC problem are identified, the engineer can evaluate the relative merits of eliminating the noise at its source, breaking the path using shielding and similar techniques, and reducing the sensitivity of the affected circuit. There are several useful resources, including publications, seminars, test labs, and consultants who specialize in solving EMC problems. The best solution is usually to begin testing a new design at the earliest possible point in the prototype phase to determine where the potential problem areas are so they can be addressed with the least cost and schedule impact.

Electrostatic Discharge Effects

Electrostatic discharge (ESD) is an important design consideration in embedded applications because of the potential for failure and erroneous operation in the presence of external electric fields. ESD voltages are commonly impressed on embedded interfaces—on the order of tens of thousands of volts—when someone walks across a floor in a low humidity environment before touching an electronic device. One of the most common places where this becomes an issue is in the keyboard or user input device, which comes in direct contact with the outside world. This effect can cause immediate damage or upset, or may cause latent failures that show up months after the ESD event. Designers most often use shielding and grounding techniques similar to those used for safety and emission reduction techniques to minimize the effects of ESD. The same resources which are available for EMC problems are also generally of use for ESD problems.

Fault Tolerance

Increasingly, fault tolerance has become a requirement in embedded systems as they find their way into applications where failure is simply unacceptable. Many hardware and software solutions have been developed to address this need.

In order to understand how to deal with these faults, we must first identify and understand the types and nature of each type of fault. Every fault can be categorized as a "hard" or "soft" fault. Hard faults cause an error that does not go away—for example, pushing reset or powering down does not result in recovery from the fault condition. Soft faults are due to transient events or, in some cases, program errors.

Self-test and diagnostic programs may be able to identify and diagnose the failure if it is not too severe. Depending upon what type of fault occurs and which device(s) are affected, it may be possible to design a system to detect the fault, possibly even isolating the location of the fault to some degree. In the event of a soft failure, it may be possible for the designer to make the system recover from the fault automatically.

A built in self test program can be written for an embedded processor that will be able to detect faults in the following types of devices:

- Processor (if the fault is not too severe)
- Memory
- ROM
- RAM
- E/EEPROM
- Peripheral devices

Note that it is difficult, if not impossible, to detect faults in the control circuits or "glue logic" in a system. Other devices, such as memories, lend themselves to diagnostic methods.

The data contents of ROM devices can be tested for errors using one or more of the following techniques:

- Parity
- Checksum
- Cyclic redundancy check (CRC)

RAM memories and the integrity of information stored in RAM by the processor can be tested for proper operation using one of the following techniques:

- Hardware error detection and correction
- Data/address pattern tests
- Data structure integrity by checking stack limits and address range validity

Additionally, the integrity of the program and proper execution sequence by the CPU can be checked using one or more of the following techniques:

- Hardware parity error detection
- Duplicate, redundant hardware and cross checking or voting
- "Watch dog" timer that operates the CPU chip's reset line
- Diagnostics that run constantly, when the CPU has nothing else to do

Hardware Development Tools

There are two general classes of hardware development tools available to the embedded developer: *passive* analysis tools which allow looking at the operation of the system, and *active* tools which allow the designer to intrude on the operation of the system while it's running (even making changes to the system's configuration and software while it is under test). The system under test is usually referred to as the "target" system, and the computer that is used to develop, edit, compile, assemble, and download the code to the target system is called the "host" system.

Passive tools include:

- logic probes to look at static logic levels and detect pulses
- oscilloscopes to look at signal waveforms
- logic analyzers, with processor specific probes
- software to assist hardware development, scope loops

Active tools include:

- In-circuit emulators (ICE) for HW/SW integration are plugged into the application circuit (the "target" system) in place of the CPU, allowing the designer to "see inside" the microcontroller, download, and execute programs selectively.

- ROM emulators (ROM ICE) allow the designer to reduce the time it takes to edit-compile-load-debug programs by replacing the program EPROM with a RAM that can be loaded quickly and easily from the host computer.

Instrumentation Issues

One of the most significant, but often ignored, problems designers must address is the proper selection and use of test instrumentation. Improper selection and application of these tools are frequently the source of much wasted time and confusion for the designer. Two common usage problems relate to the use of oscilloscope and logic analyzer probes.

A typical scope or logic analyzer is supplied with probes that might not be expected to have an effect on the observed signal or distort the data gathered. With input impedances in the megohm range and parasitic capacitances of tens of picofarads, it might seem that the test equipment would have little or no effect on the measurement, but this is definitely not the case.

There are two common causes for measurement problems: excessive ground lead inductance, and excessive capacitive loading. These things cause at the least a potential for erroneous measurements, or at worst, they can cause the circuit under test to behave differently. Two things can be done to mitigate these problems:

1) Use the shortest possible test leads, especially for the ground connection on fast logic.

2) Use high impedance probes, especially designed for high speed applications, such as high-speed FET input scope probes.

Other instrumentation problems can be caused by misinterpretation of the sampling effects in digital scopes, the lack of glitch detection in logic analyzers, and other obscure but potentially painful "learning experiences." These can only be avoided with a good understanding of the operation of the equipment in use and some practical experience.

Software Development Tools

Most of the software development tools available to the embedded system designer fall into one of the three categories: language translator, debugger, and utility programs that generally run on the host computer. Most of the available tools have been designed to run on the x86 architecture PC, and many are available as freeware, shareware, or low cost commercial products for the more common target processor architecture.

Translators:

Assembler
Compiler
Linker
Interpreter

Debugging:

Software/firmware monitors
Processor In-Circuit Emulator (ICE)
ROM ICE

Utility:

PROM Programming
Performance measurement
Execution frequency histograms

Other Specialized Design Considerations

There are several other characteristics that the embedded system designer should become at least somewhat familiar with. These include the thermal characteristics of a system and the concept of thermal resistance, power dissipation, and the effects on device temperature and reliability. Another issue of importance in portable, hand held, and remotely located systems is the application of battery power storage.

Thermal Analysis and Design

The temperature of a semiconductor device, such as a voltage regulator or even a CPU chip, is a critical system operating parameter. The reliability of these devices is also closely related to temperature, so much so because the device's reliability drops *exponentially* with increasing temperature. Fortunately, calculating the operating temperature of a device is not too difficult, as there is a simple electrical circuit analogy that is most often used to compute temperature of a device. The temperature is analogous to voltage, the power dissipated is equivalent to current, and the thermal resistance is equivalent to electrical resistance. In other words:

Temperature rise (°C) = power (watts) * thermal resistance (°C/watt)

The thermal resistance of multiple mechanical components stacked one upon the other add, just as series resistors are equivalent to a single resistor equal to the sum of the individual values.

For example: Given a 5 volt linear voltage regulator with a 9 volt input providing 1 ampere of load current, the regulator will dissipate:

P = V*I = (9–5 volts) *1 amp or 4 watts of power.

If the regulator is specified with a thermal resistance between the semiconductor junction and case of 1°C/watt (signified as Θjc), and the heat sink the regulator is mounted to has a thermal resistance from the regulator mounting surface to still ambient air of 10°C/watt (signified as Θca), then the total thermal resistance between the semiconductor junction and ambient air is:

Θja = Θjc + Θca = 1 + 10 = 11 °C/watt

The temperature rise of the junction above that of the air surrounding the regulator will then be given by:

T = P * Θja = 4 watts * 11 °C/watt = 44 °C above ambient.

If the regulator was specified to operate at a maximum junction temperature of 85°C, then the device should not be operated in ambient air of temperature higher than 85 – 44 = 41 °C, or the regulator will fail prematurely. If this is not acceptable, then the designer must reduce the input voltage to reduce the power dissipated, reduce the thermal resistance by forced air flow, or change the design to another type (e.g. a switch mode regulator) so as to keep the regulator junction within operating constraints.

Battery Powered System Design Considerations

The rapid increase in the use of portable, battery operated electronic devices has spurred the development of new battery technologies for these applications. The older single-use and rechargeable battery chemistries have been supplanted by newer ones, providing improved power densities, operating life, and other enhancements. Unfortunately, these new energy storage devices come with new and different characteristics and limitations when compared to the older energy storage devices.

Batteries are generally divided into two common groups: primary (one time discharge and discard), and secondary (rechargeable) batteries. Primary memories include the non-rechargeable alkaline and lithium cells sold commercially, and secondary cells include the older lead-acid and nickel-cadmium (NiCd) chemistries, as well as the newer nickel metal hydride (NiMH) and rechargeable alkaline and lithium ion chemistry products. There is also a wide range of special purpose batteries that are optimized for some specific characteristic, such as the zinc-air primary cell, which uses atmospheric air as an "electrode" to provide very high energy density at low operating current.

Primary batteries, such as alkalines and lithium coin cells, are relatively simple to use, but are often limited to one to three years of operation. This is primarily due to the shelf life limit imposed by internal leakage current that discharges the battery slowly over time, especially at high temperatures.

The secondary, rechargeable battery types each have slightly differing charge-discharge requirements and limitations which must be considered for effective application in a battery powered system. There are special algorithms to optimize the performance and service life of the batteries, and there are even chips which are design specifically to manage the charge and discharge of common secondary battery types.

Many embedded devices must be designed to operate for long periods of time with very little power obtained from solar cells, batteries, and other limited power sources. As a result, there are CMOS processors and memories which have been designed with very low power consumption operating modes, frequently referred to as "sleep," "power down" or "idle" modes that consume current in the μA range.

Processor Performance Metrics

In an effort to compare different types of computers, manufacturers have come up with a host of metrics to quantify processor performance. These metrics include:

The successful application of these devices in an embedded system usually hinges on the following characteristics:

- IPS (instructions per second)
- OPS (operations per second)
- FLOPS (floating point OPS)
- Benchmarks (standardized and proprietary "sample programs") that are short samples indicative of processor performance in small application programs

IPS

IPS, or the more common forms, MIPS (millions of IPS) and BIPS (billions of IPS) are commonly thrown about, but are essentially worthless marketing hype because they only describe the rate at which the fastest instruction executes on a machine. Often that instruction is the NOP instruction, so 500 MIPS may mean that the processor can do nothing 500 million times per second!

OPS

In response to the weakness in the IPS measurement, OPS (as well as MOPS and BOPS, which sound fun at least) are instruction execution times based on a mix of different instructions. The intent is to use a standard execution frequency weighted instruction mix that more accurately represents the "nominal" instruction execution time. FLOPS (megaFLOPS, gigaFLOPS, etc.) are similar, except that they weight floating-point instructions heavily to represent heavy computational applications, such as continuous simulations and finite element analysis. The problem with the OPS metric is that the resulting number is heavily dependent upon the instruction mix that is used to compute it, which may not accurately represent the intended application instruction execution frequency.

Benchmarks

Benchmarks are short, self-contained programs which perform a critical part of an application—such as a sorting algorithm—that are used to compare functionally equivalent code on different machine. The programs are run for some number of iterations, and the time is measured and compared with that of other CPUs. The weakness here is that the benchmark is not only a measure of the processor, but also of the programmer and the tools used to implement the program. As a result, the best benchmark is the one you write yourself, since it allows you to discover how efficiently the code you write will execute on a given processor with the tools available. That's as close to the real application performance as you're likely to get, short of fully implementing the application on each processor under evaluation.

Device Selection Process

In selecting a device from a field of several devices, there is more to be considered than just the speed of the processor. Some factors, such as the availability of secondary suppliers may be an absolute requirement in some applications. In order to make a systematic evaluation and selection of the best alternative, the following method has proved to be valuable, particularly when the selection process must be documented and justified. The process consists of three major steps: eliminating the alternatives that are completely inappropriate, ranking the remaining options, and evaluating the adverse consequences of a catastrophic event.

The three decision matrices are:
1) Pass/fail criteria for elimination of non-conforming alternative.
2) Weighted scoring of parametric values to rank options.
3) Consideration of adverse consequences, including their probability and severity

The first matrix consists of a table with all the options on one axis and all the "must have" criteria on the other axis. Each criterion is checked off for each option. The second matrix consists of the surviving options from the first matrix on one axis of a table, and a list of quantitative measures on the other axis, along with a weighting factor for each measure, indicating its relative importance. Each option receives a weighted score allowing them to be ranked. Finally, each of the top ranking options is evaluated with respect to probability of occurrence. For instance, a dual source part that both manufacturers produce in the Silicon Valley could become totally unavailable from either source in the event of a major earthquake in that region. In that case, even though the probability of occurrence is very low, the consequences are very severe; production could be interrupted for a very long time from both sources simultaneously, causing the product they're designed into to stop shipping for an indefinite period of time.

Other Interfaces

Many of the more advanced microcontrollers come with extensive enhancements to simplify their interface with real-world devices. There are a large number of sensors and actuators that can be interfaced to a microcontroller. Common sensors indicate parameters which include temperature, pressure, position, speed, flow rate, strain, torque, volume, density, magnetic compass heading, light level, concentrations of gases, and many more. Because of the application of semiconductor fabrication technology to many of these sensors, the cost, complexity, and accuracy have improved significantly. There are also several low cost output devices and actuators available for use with microcontrollers, including LEDs, LCDs, radio control servos, "muscle wire" that changes length when a current runs through it, and piezoelectric transducers, among others.

In many cases, these sensors and transducers inputs and outputs can be processed using simple I/O devices commonly available on most microcontrollers.

- Non-contact proximity sensors are available which put out a frequency or phase signal that is proportional to position. A simple counter can be used to measure the frequency or timing of the signals from such devices.

- A three-pin IC is available which contains all the circuitry necessary to convert the temperature into a serial digital value that can be read by a micro (Dallas DS1620).

- A simple magnetic compass provides heading information in serial digital format, or as an output voltage proportional to heading.

There are also many different options for connecting and communicating with these devices, including IR (infra-red light), radio, AC carrier current, and several variations on traditional wired connections.

Analog Signal Conversion

Many types of embedded computer applications must deal with information that is not inherently digital by nature. Real world signals, such as temperature and pressure, are inherently analog signals. Analog signals are continuously variable in amplitude and must be converted to discrete digital approximations for use in digital processors. Real analog values can only be approximated with a discrete digital value. As noted in previous chapters, devices that convert from the continuously variable form to the discrete form of representation are called *analog to digital converters* (ADC or A/D). Similarly, there are devices that convert from digital to analog form, called *digital to analog converters* (DAC or D/A). Since analog values may vary continuously over time, it is also necessary to sample these varying values to allow conversion to a single value. Sampling is like taking a "snapshot" of a changing value at one point in time, similar to the way a moving object is frozen at one point in time by a strobe light. An analog device, known as a *sample and hold* (S/H or SAH) is used to take the snapshot using a switch and a capacitor to sample and store an analog value. After an analog signal is sampled, it can be converted to digital form by an A/D converter. The digital approximation of the sampled analog value can then be used by the processor and later converted back to an analog value by a DAC, if required. This is the general approach used to record and playback speech in a digital answering machine (this will be discussed later in this chapter).

Some microcontrollers include A/D converter hardware, with as many as eight analog inputs. Most devices do not have an internal DAC, but some have a pulse width modulated (PWM) digital output instead, which can be used in place of a conventional DAC. The PWM waveform is most often generated by operating one or more of the microcontroller's timer/counters in a special PWM count mode. The PWM output has a rectangular wave output with a duty cycle that can be programmed between 0 and 100%. By averaging or integrating the PWM output with a filter, it is possible to get an analog value from this inherently digital counter output. In some cases the averaging is part of the output device's inherent characteristics. One example is an electric motor that will respond to the average value of the voltage applied to it. The rotational inertia of the motor provides the averaging of the variable duty cycle digital waveform applied to it. A resistive heating element also responds to the average level applied to it due to a relatively slow thermal time constant.

Another common form of conversion, used for digital signals, is *logic level conversion*. This is required for serial I/O devices conforming to the RS-232 standard, which uses logic voltages in the -12 to +12 volt range rather than the lower voltages that are standard on digital processors and logic. There are special level translation ICs which have voltage multipliers and negative voltage generators as well as level converters on a single IC. These devices take a +5 volt supply, convert it to + and -12 volts, and translate to and from standard logic levels. Logic level conversion is also required when interfacing two incompatible logic families, such as TTL and ECL.

Special Proprietary Synchronous Serial Interfaces

Many embedded systems require the use of a few specialized I/O devices, and the limited pin count of a microcontroller chip can make it difficult to interface all the desired I/O. In order to allow I/O expansion without using many of the pins on a microcontroller, several manufacturers have adopted a serial bus mechanism. Some of the devices are unique and proprietary, but there are two that are standardized:

- Philips' serial bus, trademarked as I²C (for Inter-Integrated Circuit bus)
- National's serial bus, trademarked as MicroWire

The I²C bus is much more flexible because it allows many devices to coexist on the bus. It is also more complex, as it allows for a large number of device addresses and multiple masters. The MicroWire bus is relatively simple, but requires additional I/O pins for multiple devices.

Unconventional Use of DRAM for Low Cost Data Storage

In some applications, static RAM (SRAM) is too expensive for data storage. A low cost alternative is to use dynamic RAM (DRAM) and handle the address multiplexing and refresh under software control. On a cost-per-bit basis, DRAM is significantly less expensive than SRAM. If the cost of address multiplexing and refresh hardware is added to the DRAM cost, it is not cost effective for small memories. In general, interfacing a DRAM directly to a microcontroller under software control is the best way to get extremely low cost-per-bit storage. It's used for applications like voice storage in low cost digital answering machines. It works well, and there are a lot of tricks you can use, such as refreshing all the rows in one burst. The disadvantage is that a significant amount of processor time has to be used to refresh the memory. In addition, each read or write access has the overhead of multiplexing the address bits and strobing the /RAS and /CAS lines under program control.

In some cases the entire memory is not needed, so it is possible to reduce the number of I/O pins used to interface to the address lines. This would seem to be wasteful, but the price of memory chips must be considered. For current chip designs, larger memories cost more than smaller ones. Once DRAM parts become obsolete, the prices for small, obsolete parts actually become greater than larger memories because the smaller chips are no longer produced in volume. It is possible to use a portion of a larger memory chip by connecting some of the address lines in parallel and ignoring the additional memory. The reason you can't just fix some of the address lines high or low is that some devices require a changing level on the address lines for internal circuitry that pre-charges the select lines in the array. The locations you can't access won't be refreshed, but that won't matter since they're not used.

Modern DRAMs have automatic refresh circuits which perform a refresh cycle using /CAS before /RAS refresh cycles, and even include internal refresh address counters. As an example, a 1Mx4 DRAM part provides 512 kilobytes of data four bits at a time. It can be fully refreshed by pulsing /CAS then /RAS low once for every row in the memory array. Having access to four bits at a time reduces the address multiplexing I/O overhead compared to using a 4Mx1 DRAM.

Digital Signal Processing / Digital Audio Recording

A common use for DRAM is in low cost digital voice recording, such as that used in some digital answering machines and toys. A microcontroller could be used in conjunction with a DRAM to record and play back voice. Standard telephone digital voice circuits sample at a rate of 8000 samples per second companded at eight bits per sample, which is 8 kilobytes/second, or 64 kilobits/second. Telephone circuits have a theoretical 4 kilohertz Nyquist bandwidth limit, but a 3 kilohertz practical audio bandwidth due to filter design constraints, which is consistent with the bandwidth of an analog phone system. At 8000 samples per second, it would only be possible to store four seconds of audio in a 32 kilobyte SRAM. Using a 1Mx4 part would allow 512/8 = 64 seconds of speech in one DRAM chip.

Standard telephone CODEC (COder/DECoder) ICs have special logarithmic analog to digital and digital to analog converters as well as low pass anti-aliasing and smoothing filters built in. They're used in huge quantity in digital telephone equipment. CODECs have serial I/O, but at 64,000 samples per second they're probably too fast for devices such as a programmable interface controller (PIC). It is also possible to reduce the sample rate if a reduced bandwidth is acceptable.

A four chip system consisting of a microcontroller, a DRAM IC, a CODEC IC, and an audio amplifier IC could be used to store and play back speech at a cost of a few dollars. The length of the recording can be increased using data compression techniques. Special compression algorithms reduce the redundancy inherent in most audio signals, such as voice. There are some very efficient coding schemes such as *linear predictive coding* (LPC) that have the ability to store compressed speech at rates as low as a few thousand bits per second. They actually model the human vocal tract. The trade off is that the computational load for compression and decompression are fairly large to get high compression ratios. It's fairly simple to playback and is useful for pre-recorded speech. That's what is used in many talking toys like Texas Instruments' "Speak and Spell." TI developed the LPC algorithm, and was first to sell it in consumer products.

Simpler compression schemes, like *adaptive differential pulse code modulation* (ADPCM), can give as much as 4:1 compression ratios without much computation. A compression ratio of 4:1 would result in 2 kilobytes per second of compressed speech. ADPCM encodes differences between samples instead of the raw values. Some applications don't require high quality audio, so there are quite a few corners that can be cut. For example, it's possible to reproduce intelligible speech using samples of less than eight bits. Four bits is probably enough for some voice storage applications. It is even possible to reproduce intelligible speech on the one bit digital output of the PC's speaker! At the other extreme, some signals, such as music, require higher sample rates and more bits per sample. Compact audio discs (standard CDs), for example, use

44,100 samples per second at 16 bits per channel per sample to store very high quality audio. This results in 44,100 samples/second * 16 bits/sample/channel * 2 channels = 1,411,200 bits per second of stereo audio. (Actual data rates are slightly higher, in order to accommodate synchronization and other overhead.)

Hardware Design Checklist

A complete and reliable design requires *all* of the innumerable details to be evaluated and analyzed correctly. The following checklist is intended to provide a guide for the designer to ensure that all the important design aspects have been evaluated. This up-front effort is a significant effort, but is less expensive and time consuming than searching for the errors once a design has been committed to production.

Schematics are an essential part of any hardware design review. To facilitate the review, here are a few general guidelines that should be followed during preparation of the schematics for a project:

- Multi-page schematics should be structured hierarchically.
- A top level sheet should be created showing the interconnects between other sheets.
- In each drawing, all inputs should on the left side and a;;outputs on the right side of the page.

Detailed Checklist

This checklist can be used as the basis of a technical design review, or in evaluating the correctness of hardware designs produced by others.

List all integrated circuits used in the design, along with the required supply voltage and percentage tolerance or range of voltages, and the actual power supply voltage range that will be encountered by these devices. Some CAD systems will assist with this process, but they may be more trouble to use than the effort warrants. Most of these analyses can be documented and calculated using a simple spreadsheet.

1. Define Power Supply Requirements

All power supply voltages and tolerances should be listed, along with the typical and maximum current requirements over the temperature range for each device. A crtitical and highly reliable design should leave significant margin (50 to 100% excess capacity) between required load and maximum available supply current, reducing the stress on the power supply. This is particularly important, since heavy loading on a power supply increases the temperature of the power handling components, significantly reducing the long term reliability of the power supply. Power supplies are among the devices in a system which are the most likely to fail, and often take other components with them when they do.

Example:

IC#	Type	Supply Voltage, %Tol., curr	Alternate Voltage(s)
U1	80C552 CPU	$Vdd = 5$ V, +/-10%, 100 mA	Vref 2.5 V +/-1%, 10 mA
U2	D/A Converter	$Vcc = 4.5$-5.5 V, 50 mA	-5 V 5%, +12V 10% 100 mA

From this data, the power supply requirements would be:

- $Vdd = Vcc = 5$ volts +/-5% 150 milliamperes minimum plus 100 milliamperes margin becomes 5 volts 5% at 250 milliamperes.

- Vref =2.5 V +/-1% 10 milliamperes minimum derived from +5 V supply using a 5 V reference IC.

- –5 V 5% 100 milliamperes minimum plus 50 milliamperes margin becomes –5V 5% at 150 milliamperes.

- +12V 10% 100 milliamperes minimum plus 50 millamperes margin becomes +12V 10% at 150 milliamperes.

Verify that the voltages delivered to all the devices are within their specifications, and that the sum of the worst case currents used by the devices can be supplied by the power source with some margin.

When a prototype circuit is available, measure actual power consumption to verify that it is within expected limits. The current consumption of subsequent units can be compared to a known good device.

2. Verify Voltage Level Compatibility

The voltage levels that will occur at the interface to each type of chip that is in the design must be compatible. This must be evaluated for two purposes: so that the correct output logic level is interpreted by the driven input, and to avoid potential damage to device inputs. The ability of the device to tolerate input voltages without damage is usually defined as an *absolute maximum rating* and the normal operating logic levels are defined in a section that is usually called *DC characteristics*. An example of the maximum levels is the Vih maximum spec, which defines the maximum input voltage that an input can withstand without potentially damaging the device's input. A 3 volt gate might have Vcc+0.3 volt maximum input specification, and driving it with the output of a 5 volt logic gate can damage the 3V gate input.

A key element of voltage level compatibility is *noise margin analysis*. Look at the Voh-Vih and Vil-Vol logic levels on all parts that interconnect to determine if sufficient noise margins are available. The hard part is determining just what an acceptable noise margin is for a given device. Several issues must be considered, including the anticipated noise environment and the required reliability level. Clearly it would be prudent to insure a high level of noise immunity designing with large noise margins for a cardiac pacemaker design! A hand-held game would not warrant the same level of reliability and resulting expense. If there is TTL compatible logic in a system, it probably doesn't make sense to design for a noise margin in excess of the inherent 400 millivolts level inherent in the TTL specs. When evaluating the noise margin of devices such as a microcontroller and an memory, it's fairly common to find that the memory's specs result in a noise margin of 200 millivolts or less, as shown in the example below:

Noise Margin Analysis - Example

OUTPUT			Vol max	Voh min	INPUT		Vil max	Vih min	Noise Margin logic zero	logic one
Signal	Pin(s)	Source			Load(s)	Signal				
CS	29	8051	0.40	2.00	EPROM	OE/	0.80	2.30	0.40	0.30
RD/	17	8051	0.40	2.00	SRAM	OE/	0.80	2.20	0.40	*-0.20*
			0.40	2.00	82C55	RD/	0.80	2.00	0.40	*0.00*

Since many systems employ logic using different power supply voltages, such as mixed 5 and 3.3 volt logic, it is important to verify that the signals that cross the boundary have sufficient noise margins, and do not exceed the maximum input voltage ratings. In some cases, level conversion or voltage clamping circuits may be necessary. Some 3 volt logic devices are tolerant of 5 volt signal levels on some of their input pins, simplifying the design. On the other hand, 3 volt CMOS outputs can often drive 5 volt logic with TTL compatible inputs directly.

3. Check DC Fan-Out: Output Current Drive vs. Loading

Maximum logic output currents (I_{OL} and I_{OH}) are specified, usually at a specific output voltage (V_{OL} and V_{OH} respectively) . The total load current that an output drives must be compared to the inputs and any resistors the output must drive, and sufficient margin must be allowed to guarantee proper operation.

Some logic outputs, such as IRQ and DMA request lines, frequently use open-drain or open-collector buses, which require pull-up resistors. Open-drain or open-collector outputs must be identified and pulled up with an appropriate resistor. Unused inputs should be pulled to their inactive state: either pulled up to the supply through a resistor, or connected to ground, as appropriate.

Pull-up resistor values must be chosen to minimize the rise time using as small a value as will satisfy the maximum I_{OL} of the weakest open-drain device driving the line.

4. AC (Capacitive) Output Drive vs. Capacitive Load and De-rating

Device timing is usually specified under specific loading conditions on the outputs. If the actual capacitive load on the outputs, consisting of the driven logic inputs and stray wiring capacitance, exceeds the load capacitor specified in the output device's timing test conditions, then the timing specs will not be valid. If the amount of overload is not severe, it is possible to estimate the additional delay required to charge the excess capacitance. The delay depends upon the available charging current and actual load capacitance.

DC and AC loading can be summarized in a spreadsheet as shown below:

Source						Load				Unit Load			Total		
			uA	uA	pF					uA	uA	pF	uA	uA	pF
Signal	Pin#	Source	IOL	IOH	CL	Load	Signal	Qty		IIL	IIH	Cin	IIL	IIH	Cin
AD0..7	39-2	8051	3200	-800	100	74LS373	A0..7	1		-400	20	10	-400	20	10
(P0.0-P0.7)						SRAM	D0..7	1		-1	1	7	-1	1	7
						EPROM	D0..7	1		-1	1	12	-1	1	12
						82C55	D0..7	1		-10	10	20	-10	10	20
						wire cap		5				2			10
											Total		-412	32	59
											Margin		2788	768	41
		SRAM	1600	-600	50	74LS373	A0..7	1		-400	20	10	-400	20	10
						8051	D0..7	1		-1	1	20	-1	1	20
						EPROM	D0..7	1		-1	1	12	-1	1	12
						82C55	D0..7	1		-10	10	20	-10	10	20
						wire cap		5				2			10
											Total		-412	32	72
											Margin		1188	568	-22

5. Verify Worst Case Timing Conditions

All timing specifications should be evaluated for potential timing violations, as covered in chapter 6. This is particularly important for signals that are heavily loaded requiring de-rating of the timing specs, or tri-state signals that are subject to bus contention problems.

6. Determine if Transmission Line Termination is Required

The signal rise time and maximum trace length must be evaluated to determine if a signal interconnect must be treated as a transmission line, requiring constant impedance along the length of the trace, and termination to prevent reflections. If the signal has a fast rise time and trace length, L, greater than about one-sixth the edge length of the pulse, then it is necessary to analyze the circuit as a transmission line using this formula:

$L = T_r / D$ where

L = length of rising or falling edge in inches (in)

T_r = rise time in picoseconds (pS)

D = delay in picoseconds per inch (pS/in)

For traces on a standard printed circuit board, the value for D will be in the range of 100 to 200 pS/in. Depending upon how much distortion you're willing to live with, the critical trace length will be between one-sixth and one-quarter of the length of a trace corresponding to the signal's transition. For a trace that is shorter than one-sixth the length of the signal's rising or falling edge, the circuit seldom needs to be considered to be a transmission line. Traces that are much longer than one-quarter the length of the fastest edge will start to behave as transmission lines, exhibiting reflections of the signal when the transition gets to the far end of the trace and is reflected back to the near end. Once the trace is about half of the length it takes for a logic transition to propagate, the problems become quite pronounced.

7. Clock Distribution

Distribution of clock signals must be done in a way that compromises the need to minimize clock skew, while avoiding reflections that can cause unacceptable clock transitions due to transmission line effects. Distributing clocks in such a way as to avoid excessive skew implies the use of a clock tree to provide equal time delay to each load. However, a tree topology is in direct conflict with the need to maintain a single, stubless transmission line. The ideal transmission line is essentially "daisy-chained" with a trace that has constant impedance across its length and has no stubs, but that usually results in maximum timing skew! Clock signals should also be isolated from other signals to prevent crosstalk between the clock and other signals. Clock signals should generally NOT be gated, to avoid undesirable side effects.

8. Power and Ground Distribution

Ground and power planes are recommended on printed circuits wherever possible, because they allow low impedance connections and provide high frequency decoupling from inter-plane capacitance. Ground connections should be as short as possible, especially for ground pins on multiple output logic devices, to prevent ground bounce.

Capacitors for Bypassing Power Supply Noise

The power and ground pins of every IC should be bypassed using a capacitor with low impedance at the frequencies of interest (determined by rise time, not clock rate). The self-resonance of larger capacitors, such as 0.1 microfarad, may result in little effect on the fast current transients present in high-speed logic chips. 0.01 or 0.001microfarad (or even hundreds of picofarads) low inductance capacitors, are more appropriate for fast logic devices having sub-5 nanonsecond rise times. Multi-layer ceramic dielectric surface mount capacitors work better than leaded, tantalum or electrolytic capacitors at high frequencies. Each board in a system should also have a larger tantalum or electrolytic capacitor to provide medium frequency bypassing for peak currents.

When possible, power supply and ground connections should be made independently to the power supply, to minimize common impedances, also known as *ground loops*. This is especially important for circuits containing mixed analog and digital circuitry.

Mixed Analog and Digital Circuitry

The analog power supply should be separately regulated from the digital supply, to provide a quiet power source to the analog circuitry. Separate power and ground planes should be maintained to minimize coupling between noisy digital circuits and sensitive analog or RF (radio frequency) circuits. Analog power planes should not overlap with digital planes, as the digital noise will couple through the inter-plane capacitance. Digital and analog grounds should only be interconnected at one point, usually very near the analog-digital conversion IC.

High impedance analog signals should be physically and electrically isolated from digital signals to minimize digital noise on the analog signals.

Digital inputs that are driven by analog circuitry should be clamped, using a series resistor and low forward voltage Schottky diodes, to power and ground to clamp the signals to levels that are within the logic input specification levels.

Safety

High voltage conductors should be physically and electrically isolated from low level and user accessible signals to avoid potential shock hazards. All conductors should be sized large enough to allow carrying maximum current, under short circuit conditions, and protective devices, such as fuses and PTC switches, should be used to prevent. Conductors carrying more than 40 volts and telephone line conductors must be isolated by at least one-quarter inch from other conductors or transformer isolated for safety agency and telecom approvals.

9. Asynchronous Inputs

Asynchronous inputs should be synchronized using two levels of flip-flops to minimize the probability of a metastable state when asynchronous inputs are sampled. This is particularly important for programmable logic devices, which may have slow recovery times from metastable states.

10. Guarantee Power-On Reset State

Verify that any devices, such as CPU, PLDs, and registers, are reset to a known state when power is applied, or whenever power falls below normal operating levels (brown out condition). All CPUs, counters, registers, shift registers and memory devices are subject to unpredictable behavior when the power is out of spec and must be reset after the power returns to specified levels.

11. Programmable Logic Devices

Verify that all flip-flops in the device will be in a known state upon power-up, and that any counters and state machines with unused states will transition to a valid state in the event that they get into an invalid state.

Leave a few available input and output pins available to facilitate changes in the event that additional logic functions become necessary.

12. Deactivate Interrupt and Other Requests on Power-Up

Interrupt, DMA, and other edge sensitive input requests should be disabled upon power up to minimize the chance that a spurious event will be processed when the system is turned on.

13. Electromagnetic Compatibility Issues

Signals that enter and leave the printed circuit boards should be filtered to reduce the unintentional emission of radio frequencies as much as possible. Digital circuits should also be packaged in conductive enclosures when possible to minimize the digital signals from being radiated as electromagnetic interference to other devices, and to protect the device from external electromagnetic fields and static discharge.

High order harmonics from clock edges can be mitigated by the use of ferrite beads (small value inductors) that reduce the amplitude of the higher clock harmonics. Clocks should also be kept away from I/O signals and connectors to reduce the coupling of clock noise to wires and interconnects that can act as antennas, conducting and radiating clock harmonics as radio interference.

14. Manufacturing and Test Issues

Manufacturing of boards can be made simpler if the design implements a method that allows programming processors, memories, and PLDs while the components are mounted to the card. This facilitates manufacturing the boards prior to programming the devices. This also facilitates loading test programs into the board to allow more effective tests to determine of the board is operating as intended.

Signals which control or enable outputs or programming signals that might need to be disabled and driven externally for test purposes should be isolated from a test point with a series resistor, allowing an external test or programming circuit to drive the signal without damaging the output device on the board.

The inclusion of easily probed test points also makes it easier to diagnose failures by making it easier to probe critical signals on the board.

References, Web Links, and Other Sources

Since he number of information sources that may be of interest is too great to include a comprehensive list—and many links to the information become obsolete—the sources noted in this chapter are just the starting points for more detailed information. Some of the books listed here relate directly to this subject, and others are some of my personal favorites, as they contain information which I make reference to regularly.

An important thing to keep in mind for any source of information is who the source is and how they derive their income. Trade magazines are useful, and because they are free to qualified subscribers, they are very popular source of information. Unfortunately, they derive their income solely from their advertisers, and most of the articles are written by advertisers and the magazine editors. As a result, they often portray a very biased view of what's going on in the industry. Likewise, web sites and other advertiser supported information sources often have very slanted versions of reality. There are a few exceptions, such as magazines that are supported by subscriptions as well as advertising, that have articles written by those of us who are down in the trenches. They often provide a more accurate, though still biased, view of what's really going on.

Books

The Art of Electronics, by Horowitz and Hill, also the accompanying *Student Manual*, by Hayes and Horowitz, to accompany the text. This is an all-time favorite tome that covers an incredibly wide range of topics in a very readable and useful way. The student guide provides a refreshing review of the practical side of electronics, and will be invaluable for those who need to learn more about electronics.

The Circuit Designer's Companion, by Tim Williams is a good reference for understanding the differences between ideal circuits you learn about in school, and the things that happen in the real world. Includes a lot of material on undesirable component behaviors that the manufacturers frequently gloss over if they deal with them at all.

High-Speed Digital Design, a Handbook of Black Magic, by Howard W. Johnson and Martin Graham, which in spite of it's subtitle, is soundly based in math and scientific principles, and provides a clear description of what really happens in high-speed circuits. This is an excellent text to understand the design of reliable high-speed circuits, which often exhibit non-ideal characteristics.

The Microcontroller Idea Book, by Jan Axelson uses the 8051BASIC chip to illustrate a range of introductory embedded applications. Jan is an excellent writer, as well as thorough and practical, so you should probably just give in and go buy all of her books.

Serial Port Complete, by Jan Axelson covers use of the PC's serial port and can be very useful when interfacing an embedded controller to a PC's serial COM port.

Parallel Port Complete, by Jan Axelson covers use of the PC's parallel port and can be very useful when interfacing an embedded controller to a PC's parallel port.

Printed Circuits Handbook, by Clyde F. Coombs is the standard reference text covering the design and manufacture of printed circuit boards.

The Cartoon Guide to Physics, by Gonick and Huffman is a great introduction to physics and basic electronics, using humorous cartoons to illustrate basic principles without resorting to complex math.

A Whack On The Side Of The Head, by Roger von Oech, is a humorous and effective book describing how to <u>learn</u> to be innovative.

Web and FTP Sites

The sites listed below can be reached through links provided on the companion CD-ROM, but they can quickly become obsolete, so they are also on the book web site at www.hte.com/echdbook. In addition, the LLH Technology Publishing web site will carry updates and corrections to this book; be sure to visit them at www.LLH-Publishing.com.

Embedded Computer Engineering. The web site for embedded classes we teach at UCSD extension is: www.hte.com/uconline

Embedded Computer Hardware Design. This is the class that this book was originally created for: www.hte.com/uconline/ecd

Miller-Freeman Publishing's Embedded Web Site. This site is hosted by the publisher of the trade magazine "Embedded Systems Programming." This web site has some useful technical information, but you have to work to find it, as it's buried under a lot of advertising. www.embedded.com

Periodicals: Subscription

Circuit Cellar Ink, published monthly, covers embedded systems topics with practical, design oriented articles that often include schematics and code for working projects. This magazine leans to the practical, hands-on side of design, including the sorts of things like single chip microcontrollers that make traditional computer scientist types sputter uncontrollably.

FORTH Dimensions. This is the bi-monthly newsletter of the Forth Interest Group, and covers Forth, a very unique language. Forth is a very different and yet powerful language which is very well adapted to the embedded computing environment. This is the sort of thing that can turn a politically correct computer scientist absolutley apoplectic. On the other hand, I've never met anyone who really understood the language that didn't like it! Some people would characterize Forth fanatics as religious, but I'd say they're just sensitive because they understand the capabilities of the language and are frustrated by the common view that Forth is not an appropriate language. If you like a good fight, just yell "Forth" into a room full of Forth advocates and computer scientists!

Microcomputer Journal, Midnight Engineer, and *Robotics Digest*, all published whenever Bill Gates gets around to it. (No, he's not that Bill Gates!) This fellow is a really efficent, one-man publishing empire who does everything, including printing and binding the magazines himself. He uses his knowledge of embedded systems to help automate the publishing process. Lots of practical information in these, though the publications probably won't outlive Bill.

Periodicals: Advertiser Supported Trade Magazines

EDN Magazine, an advertising supported trade publication, covers embedded computing and general electrical engineering topics. Every September they publish a directory of microprocessors and microcontrollers that is a very useful source of information on the incredible number of devices that's out there. They also have a web site with all of their articles and other useful information at www.ednmag.com

Electronic Engineering Times, is a newspaper-like weekly trade journal which covers all EE topics including embedded systems.

Embedded Systems Programming, published monthly, covers the software aspects of embedded systems. This magazine leans to the high end and embedded x86 PC software market, and is dominated by the high-level language computer science types.

Electronic Design, a monthly EE oriented magazine is similar to EDN but with less coverage of embedded topics.

Embedded Technology™ Series

Programming Microcontrollers in C, Second Edition
by Ted Van Sickle

INCLUDES WINDOWS 95/98 CD-ROM. Completely updated new edition of a classic for embedded systems designers and programmers. It covers C basics, advanced C topics, microcontroller basics and usage, and gives example code, using the Motorola family of microcontrollers, including RISC machines. The CD-ROM contains the code from the book, a full set of Motorola's microcontroller documentation in PDF format, and a fully searchable electronic version of the text.

1-878707-57-4 $59.95

Embedded Controller Hardware Design

by Ken Arnold

INCLUDES WINDOWS CD-ROM. This practical tutorial introduces the reader to the design of embedded microprocessor- and microcontroller-based systems. General topics covered in the book include device architecture, interfacing, timing, memory, I/O, as well as design and development techniques. The book presents the latest application-oriented information concerning this rapidly changing area of technology.

1-878707-52-3 $49.95

Controlling the World with Your PC
by Paul Bergsman

INCLUDES PC DISK. A wealth of circuits and programs that you can use to control the real world. Connect to the parallel printer port of your PC and monitor fluid levels, control stepper motors, turn appliances on and off, and much more. The accompanying disk for PCs contains all the software files in ready-to-use form. All schematics have been fully tested. Great for embedded systems engineers, as well as students and scientists.

1-878707-15-9 $35.00

The Forrest Mims Engineers Notebook
by Forrest Mims III

Revised edition of a classic by world's best-selling electronics author. Hundreds of useful circuits built from ICs and other parts.

1-878707-03-5 $19.95

The Forrest Mims Circuit Scrapbook, Volumes I and II
by Forrest Mims III

More "greatest hits" circuit designs from Forrest Mims. Volume I contains digital PLLs, interval timers, light wave communicators, and much more. Volume II contains comparators, data loggers, laser diode devices, fiber optic sensors, power supplies, and much more.

Vol. I: 1-878707-48-5 $19.95
Vol. II: 1-878707-49-3 $24.95

The Integrated Circuit Hobbyist's Handbook
by Thomas R. Powers

This comprehensive "cookbook" of circuit applications is conveniently cross-indexed by device and application. Contains amplifiers, filters, bus transceivers and bus buffers for digital interfacing, counters, comparators, FSK modulators and decoders, oscillators, and much more.

1-878707-12-4 $19.95

Simple, Low-Cost Electronics Projects
by Fred Blechman

Contains a wealth of fully tested electronics design projects using commonly available parts, each with circuit theory, parts lists, and design and testing guidelines.

1-878707-46-9 $19.95

Visit our web site for more great technical books on all subjects!

LLH Technology Publishing   www.LLH-Publishing.com